GENERAL DYNAMICS
F-111 AARDVARK

By Don Logan

ROCKWELL B-1B: SAC's LAST BOMBER

NORTHROP's T-38 TALON

NORTHROP's YF-17 COBRA

THE 388TH TACTICAL FIGHTER WING AT KORAT RTAFB 1972

REPUBLIC's A-10 THUNDERBOLT II

BOEING C-135 SERIES

GENERAL DYNAMICS F-111 AARDVARK

GENERAL DYNAMICS
F-111
AARDVARK

Don Logan

Schiffer Military History
Atglen, PA

ACKNOWLEDGEMENTS

I would like to thank the following individuals for their help in the research for this book: Steven R, Hyre, whose research and data files on individual F-111 aircraft histories provided most of the serial number historical data used in this book; Mike Moore and C. Roger Cripliver of Lockheed Martin Tactical Aircraft Systems supplying contractor photos and production/test information; former F-111 flyers and Desert Storm Veterans Jim Rotramel and Craig "Quizmo" Brown supplying operational data; Chris McWilliams supplying information on the FB-111A and their nose art; and from Australia – Dave Riddel, Lenn Bayliss, and Nigel Pittaway supplying information on both USAF and RAAF Aardvarks. I also would like to thank my former F-111A Flight Commander Phil Brandt for his help proofing the manuscript and supplying suggestions for improvement of the text. Thanks to Roger Johansen for his work on the line art. The flight suit patches were supplied by Craig Brown and John Cook.

I would also like to thank the following individuals who provided photographs included in this book: Don Abrahamson, APA-AVN Photo Australia, Joe Arnold, Lenn Bayliss, Jim Benson, G.J. Booma, Tom Brewer, Craig "Quizmo" Brown, Dave Brown, Ken Buchanan, Mike Campbell, Tony Cassanova, Andrew H. Cline, George Cockle, John Cook, Harley Copic, C. Roger Cripliver, Bill Curry, Rolf Flinzner, Kevin Foy, Michael France, Alec Fushi, Jerry Geer, Jim Geer, Jim Goodall, Norris Graser, Bob Greby, Mike Grove, Paul Hart, Marty Isham, F.A. Jackson, Tom Kaminski, Craig Kaston, Duane Kasulka, Ben Knowles, Robert Lawson, Ray Leader, Nate Leong, Roy Lock, Lockheed Martin Tactical Aircraft Systems, Terry Love, Bill Malerba, Gerry Markgraf, Pat Martin, Charles B. Mayer, Chris Mayer, Don McGarry, Gerald McMasters, Chris McWilliams, Dave Meehan, Dave Menard, Stephen Miller, Mike Moore, NASA, John Owen, Kevin Patrick, Bob Pickett, Doug Remington, Geoff Rhodes, Dave Riddel, Chuck Robbins, Brian C. Rogers, Rome Air Development Center, Mick Roth, Jim Rotramel, Tony Sacketos, G. Salerno, Bill Schell, Bob Shane, Douglas Slowiak/Vortex Photographics, Keith Snyder, Bill Strandberg, Don Sutherland, Keith Svendsen, Bob Trimble, Charles Trump, Ted Van Geffen, Adrian Walker, Wayne Whited, Pete Wilson, and Scott Wilson, and 82nd Wing Photographic RAAF.

THE AUTHOR

After graduating from California State University-Northridge with a BA degree in History, Don Logan joined the USAF in August of 1969. He flew as an F-4E Weapon Systems Officer (WSO), stationed at Korat RTAFB in Thailand, flying 133 combat missions over North Vietnam, South Vietnam, and Laos before being shot down over North Vietnam on July 5, 1972. He spent nine months as a POW in Hanoi North Vietnam. As a result of missions flown in Southeast Asia, he received the Distinguished Flying Cross, the Air Medal with twelve oak leaf clusters, and the Purple Heart. After his return to the U.S., he was assigned to Nellis AFB where he flew as a rightseater in the F-111A. He left the Air Force at the end of February 1977.

In March of 1977 Don went to work for North American Aircraft Division of Rockwell International, in Los Angeles, as a Flight Manual writer on the B-1A program. He was later made Editor of the Flight Manuals for B-1A #3 and B-1A #4. Following the cancellation of the B-1A production, he went to work for Northrop Aircraft as a fire control and ECM systems maintenance manual writer on the F-5 program.

In October of 1978 he started his employment at Boeing in Wichita, Kansas as a Flight Manual/Weapon Delivery manual writer on the B-52 OAS/CMI (Offensive Avionics System/Cruise Missile Integration) program. He is presently the editor for Boeing's B-52 Flight and Weapon Delivery manuals, and Boeing North American B-1B OSO/DSO Flight Manuals and Weapon Delivery Manuals.

Don Logan is also the author of *Rockwell B-1B: SAC's Last Bomber, The 388th Tactical Fighter Wing; At Korat Royal Thai Air Force Base 1972, Northrop's T-38 Talon, Northrop's YF-17 Cobra, Republic's A-10 Thunderbolt II,* and *The Boeing C-135 Series: Stratotanker, Stratolifter, and Other Variants*. (All available from Schiffer Publishing Ltd.)

Book Design by Robert Biondi.

Copyright © 1998 by Don Logan.
Library of Congress Catalog Number: 98-84262.

All rights reserved. No part of this work may be reproduced or used in any forms or by any means – graphic, electronic or mechanical, including photocopying or information storage and retrieval systems – without written permission from the copyright holder.

Printed in China.
ISBN: 0-7643-0587-5

We are interested in hearing from authors with book ideas on related topics.

Published by Schiffer Publishing Ltd.
4880 Lower Valley Road
Atglen, PA 19310 USA
Phone: (610) 593-1777
FAX: (610) 593-2002
E-mail: Schifferbk@aol.com.
Please write for a free catalog.
This book may be purchased from the publisher.
Please include $3.95 postage.
Try your bookstore first.

Contents

CHAPTER 1: TFX PROGRAM – THE F-111 7
Program Design Requirements 7
Flight Testing 9
Australian and British F-111s 9
RF-111 Configurations 9
Total F-111 Aircraft Production 12
The General Dynamics, Fort Worth F-111 Restoration Facility (Aardvark Hospital) 13

CHAPTER 2: THE F-111 DESIGN 14
Description 15
Bombing-Navigation Systems 26
Avionics Modernization Upgrade Program (F-111A/E) and Pacer Strike (F-111F) 28
Pave Tack/Data Link System (F-111F and F-111C) 29
Pave Mover Program 31

CHAPTER 3: U.S. AIR FORCE TACTICAL F-111s 32
F-111A 32
EF-111A 89
F-111D 106
F-111E 137
F-111F 169
F-111G 206

CHAPTER 4: U.S. AIR FORCE STRATEGIC F-111s 215
FB-111A 215
Other F-111 Strategic Bomber Designs 247

CHAPTER 5: U.S. NAVY F-111s 252
F-111B 252

CHAPTER 6: ROYAL AUSTRALIAN AIR FORCE F-111s 261

CHAPTER 7: BRITISH ROYAL AIR FORCE F-111Ks 278

CHAPTER 8: COMBAT OPERATIONS 282
   Combat Lancer 282
   Constant Guard/Linebacker 283
   El Dorado Canyon 285
   Desert Shield/Desert Storm 286
   Operation Proven Force 287
   Operation Provide Comfort 288

CHAPTER 9: NATIONAL AERONAUTICS AND SPACE ADMINISTRATION (NASA) F-111s 289
   F-111 Developmental Testing 289
   Integrated Propulsion Control System (IPCS) 289
   Transonic Aircraft Technology (TACT) and Advanced Fighter Technology Integration (AFTI) 290

APPENDICES
   Appendix 1: Insignia 295
   Appendix 2: External Differences 301
   Appendix 3: Avionics Systems 302
   Appendix 4: F-111 Specifications 302
   Appendix 5: F-111 Engine Specifications 303
   Appendix 6: F-111 Unit and Tail Code Summary 304
   Appendix 7: Attritted F-111 Aircraft 307
   Appendix 8: F-111 Aircrew Fatalities Worldwide 1967-1998 310
   Appendix 9: F-111 Aircraft on Display 314

(Author)

CHAPTER ONE

TFX Program - The F-111

PROGRAM DESIGN REQUIREMENTS

The TFX, later designated the F-111, the last of the century series fighters, was designed to fill the need for a new fighter bomber. The requirements called for an aircraft with the payload of a B-66 (15,000 pounds of bombs) and the supersonic speed and agility of the F-105.

The General Operational Requirements (GOR) were issued on March 27, 1958. The GOR called for Weapon System 649C- a 1964 Tactical Air Command Mach 2+, 60,000 foot maximum altitude, all-weather fighter, capable of vertical and short takeoff and landing (V/STOL). The Air Force canceled this 1958, GOR (No. 169) on March 29, 1959, due to the fact that operational vertical takeoff capability was not yet feasible. The GOR was replaced by a System Development Requirement (SDR No. 17) on February 5, 1960. This SDR encompassed most of the requirements of the canceled GOR, with the vertical takeoff and landing (VTOL) requirement deleted. Combined with the Tactical Air Command's (TAC's) revised specifications and a delayed first operational date, the new designation of the aircraft was Weapon System (WS) 324A.

A Specific Operational Requirements (SOR 183) document was issued on July 14, 1960, as a follow-on to SDR 17. It further defined Weapon System 324A as air superiority, Mach 2.5, 60,000-foot-plus altitude, all-weather, day and night, two-man crew, STOL fighter (that could take off or land, even on sod fields, in less than 3,000 feet), with an 800-mile low-level combat radius (including 400 miles close to the terrain at Mach 1.2 speed), and carrying either conventional or nuclear weapons. The unrefueled 3,300-nautical mile ferry range and 1,000-lb internal payload, and lifting payload between 15,000 and 30,000 pounds required by SOR 183, put the design in the fighter-bomber class. The aircraft resulting from this design was expected to begin replacing the TAC F-100Ds and F-105s in 1966. Previous fighters had been designed to climb and maneuver rapidly, but could not carry a large payload and had a relatively short range. The weapon system design of SOR 183 called for a versatile aircraft with good payload and range which could act as a strike/attack fighter, a tactical fighter, or an interceptor.

The U.S. Air Force decided that a variable-sweep wing design with an improved afterburning turbofan engine would satisfy SOR 183. Air Force requirements under WS 324A included a reconnaissance version, procuring enough aircraft to equip six squadrons and enough fighter bombers to equip a minimum of six tactical wings.

During October 1960, the U.S. Air Force prepared to inform the defense industry of its new fighter requirement with a Request For Proposal (RFP). The Office of the Secretary of Defense (OSD) asked in November 1960, that the RFPs requested in October be withheld for further review of SOR 183. In December the project was given a new name, the Tactical Fighter Experimental (TFX).

The October RFPs remained on hold until February 16, 1961, when the new Secretary of Defense, Robert S. McNamara, asked the Air Force to determine with the Army and Navy if the TFX could provide close air support (CAS) to ground troops, air defense of the fleet, as well as the Air Force's primary objective, interdiction of enemy logistics. McNamara believed a single tri-service fighter for the Air Force, Navy, and U.S. Army would save money. The Army and Navy wanted a simpler CAS airplane, preferably the Navy-sponsored VAX (Attack Aircraft, Experimental). The Air Force did not go along with the VFX, but did agree that the TFX was not the aircraft for close air support. Army and Navy CAS objections to the TFX finally prevailed in May 1961, but McNamara remained convinced that the TFX could satisfy other Navy and Air Force needs. In June he instructed the Air Force to "work closely" with the Navy in tying the two services' requirements into a new, cost-effective TFX configuration.

The go-ahead decision by McNamara was accompanied by a revised SOR 183, reflecting his arbitration of Air Force and Navy unreconciled requirements. In spite of these unreconciled requirements, Admiral George Anderson, Chief of Naval Operations, and General Curtis LeMay, Air Force Chief of Staff, with the Secretary of Defense's approval, publicly announced their endorsement of a new tactical fighter program on September 1, 1961.

The September 1961, modified SOR 183 called for a wider fuselage (to satisfy Navy needs for more internal fuel and a panoramic nose antenna), with overall dimensions and weight kept to the maximum acceptable for carrier operation. The Air Force's TFX version could have a gross takeoff weight of 60,000 pounds (20,000 less than anticipated), compared to the Navy's 55,000.

New RFPs, replacing the October 1960, RFPs were issued on September 29, 1961. They were sent to Boeing, Chance-Vought, Douglas, General Dynamics, Grumman, Lockheed, McDonnell, North American, Northrop, and Republic. Only Northrop declined the USAF invitation for proposal, and responses from the remaining nine companies were received in early December. The Air Force Selection Board and a Navy representative endorsed the Boeing proposal on January 19, 1962, but the Air Force Council rejected it. In late January the Air Force and Navy agreed that none of the contractor proposals were acceptable, but that both the Boeing and the General Dynamics designs deserved further study. A $1 million contract for more design data was issued to both companies in February 1962. Meanwhile, the bi-service TFX was renamed. The Air Force's future version of the TFX was designated F-111A; the Navy's, F-111B. In May both the Air Force and Navy Secretaries disapproved the two contractors' second proposals for lack of sufficient data. Third proposals appraised in late June brought another impasse. The Air Force endorsed the Boeing input, but the Navy "refused to commit ... unequivocally with this program until after the design had been defined." McNamara on July 1, 1962, ordered a final runoff on the basis of open "pay-off points" for performance, cost, and commonality. After receiving an additional $2.5 million apiece, Boeing and General Dynamics submitted their fourth and last proposals in September. The Air Force Selection Board as well as the Air Force Council again chose the Boeing design, but on November 24th the Secretary, possibly influenced by the Texas Congressional delegation and Vice President Lyndon Johnson, publicly ruled in favor of General Dynamics.

On December 21, 1962, the Air Force initiated procurement of 23 Research, Development, Test, and Evaluation (RDT&E) F-111s (18 F-111As and 5 F-111Bs) by amending the Letter of Contract which was initially issued in February 1962, which had covered General Dynamics' second competitive proposal. The Air Force wanted to get production started without awaiting negotiation on a definitive contract. The $28 million December 1962, amendment to the contract made possible urgent subcontracts and in November 1963, an agreement with Grumman (the number one subcontractor of the General Dynamics team because of their experience with the manufacture carrier based fighter aircraft) for development and production of the Navy version, the F-111B.

Twenty-two million dollars plus was obligated to the Navy for development and hardware of a Pratt & Whitney TF30 engine. The Air Force assumed this Navy responsibility in late 1967, after the TF30 had undergone several transformations. By the spring of 1964 AiResearch, AVCO, Bendix, Collins Radio, Dalmo Victor, General Electric, Hamilton Standard, Litton Systems, McDonnell Aircraft, Texas Instruments,

474th TFW F-111As 66-0033 (442nd TFTS) and 67-0044 (428th TFS) join up over the Nevada desert (Author)

and seven other major subcontractors had become involved in the F-111 program, doing business with 6,703 suppliers in 44 states. The Hughes Aircraft Company, an associate prime contractor for the F-111B's Phoenix missile system, had also signed a contract with the Navy.

On May 1, 1964, the amended Letter of Contract of February 1962 was finalized as a fixed price incentive fee (FPIF) contract (AF 33-657-8260), with a 90/10 percent sharing arrangement. The ceiling price ($529 million) was based on 120% of the $480.4 million target cost for the 23 RDT&E F-111s. This included flight testing, spares, ground equipment, training devices, static and fatigue test data. The FPIF development contract of May 1964 contained cost, schedule, performance, and operational clauses, plus a provision for the "correction of deficiencies."

A rise in anticipated cost from an estimated $4.5 to $6.03 million per aircraft caused the DoD, in early 1965, to cut back the F-111 program sharply. Additional USAF requirements which included improved avionics and requirements for a strategic F-111 bomber to replace B-52C through B-52F aircraft, shaped the future of the program. Development of a reconnaissance F-111 which was approved in October 1965, (but eventually canceled) was also a factor in the increased cost.

The Air Force started procurement of the F-111 production aircraft as it had the RDT&E aircraft. It publicly announced on April 1965, that it had awarded General Dynamics a fixed price incentive fee contract, authorizing the production of 431 F-111s. This was more than a 50-percent reduction of the total aircraft originally planned. The production contract also authorized negotiation of an unusually large number of subcontracts. These subcontracts were mostly with firms already involved in the F-111 development.

The production contract of 1965, was replaced by a multi-year, Fixed Price Incentive Firm contract (AF 35-657-13403) in May 1967. Production was then raised to a total of 493 F-111s – 24 Navy F-111Bs (later, all but two were canceled); 24 F-111Cs for Australia; and 445 F-111s (including 50 first programmed for the RAF) for the U.S. Air Force.

By mid-1973, the Air Force had accepted 533 of a future grand total of 563 F-111s. The 563 were made up of; 158 F-111As (18 were preproduction RDT&E aircraft and 10 were production aircraft dedicated to testing); 7 F-111Bs for the Navy (5 RDT&E and 2 production units); 24 Australian F-111Cs; 2 F-111Ks (built but not completed as part of the British order which was later canceled); 94 F-111Es; 96 F-111Ds; 76 FB-111A medium-range strategic bombers; and 106 F-111Fs (of which 1 was destroyed before delivery).

After 1970, the Congress insisted on funding 48 additional late model F-111Fs. The Air Force, however, had bought just 36 and was able to defer acquisition of the remaining 12 which were regarded as not needed. The total cost of the F-111 series was $5.479 billion for 541 F-111s (excluding the 23 RDT&E F-111s, 18 for the Air Force and 5 for the Navy, but included the single F-111F destroyed before delivery). The contractor's lower figure ($5.431 billion) still represented an overall target cost increase of $3.228 billion.

FLIGHT TESTING

Testing was accomplished in three phases: Category I, Category II, and Category III.

Category I Testing was accomplished by the contractor. It included all basic tests required to prove that the airframe, power plant and all associated subsystems met contract specifications.

Category II Tests were conducted by the Air Force Systems Command (AFSC) before the aircraft was turned over to the using command (Tactical Air Command – TAC for the F-111A, F-111D, F-111E, and F-111F; and Strategic Air Command – SAC for the FB-111A). During this period Air Force pilots, and in the case of the F-111, NASA pilots, flew the aircraft and checked equipment to determine how well they met combat flying stresses and requirements.

Category III Test program included testing and evaluation of the aircraft and its related equipment under operational conditions by the using command. This category of testing was also used to develop tactics and techniques and to determine any operational restrictions. During this phase the logistics support requirements were established and personnel proficiency rating were determined.

AUSTRALIAN AND BRITISH F-111s

The F-111C, a modified F-111A, was specifically designed for the Royal Australian Air Force (RAAF). Modifications included the longer FB-111A wings and a heavier landing gear (similar to that of the FB-111A). The RAAF bought 24 F-111Cs, later supplementing them with four F-111As and fifteen F-111Gs. The F-111As were modified by the Australians to F-111C standard. (See Royal Australian Air Force F-111s chapter).

The British Royal Air Force also contracted for versions of the F-111, the F-111K. It was an F-111A featuring more advanced avionics and the FB-111A's landing gear. Two of 50 programmed F-111Ks were built, but not completed and were never flown. They were disassembled following Great Britain's cancellation of its order in January 1968. (See British Royal Air Force F-111K chapter).

RF-111 CONFIGURATIONS

The requirement for development of a reconnaissance version of the F-111 originated in the first TFX proposal of September 29, 1961, which called for 876 aircraft, including 110 to equip six squadrons for tactical reconnaissance. Two versions were planned, one for the USAF based on the F-111A, and one for the U.S. Navy based on the F-111B. The numbers required changed from the 110 in 1961, to 305 in June 1963, to 60 RF-111Ds USAF and six RAAF RF-111Cs during the last quarter of 1967. In 1969, the USAF changed once again, now to converting 46 F-111As already completed

This RAAF F-111C is carrying wall to wall MK 82s, 48 total. Though the aircraft is capable of carrying this load, it is rarely done. The two outboard pylons on each wing do not pivot, and as a result, the wings must remain in their 26 degree forward position, limiting the Vark's top speed. (82nd Wing Photographic, RAAF)

to RF-111A. By July 1970, the USAF no longer showed any interest in the RF-111 program.

In October 1965, approval for a single RF-111A prototype airplane was given by DoD. On March 28, 1966, F-111A No. 11 (63-9776) was designated as the prototype for the reconnaissance version of the F-111. The aircraft, still on the assembly line, was accepted in "as is" condition and assigned to the RF-111A program. The main modifications required to produce the RF-111A were removal of the weapons bay doors and the installation of a pallet in the bay to house the various reconnaissance sensors and related reconnaissance equipment.

On August 10, 1966, the RF-111A mock-up was viewed by representatives of Tactical Air Command (TAC), Tactical Air Reconnaissance Center (TARC, Shaw AFB, South Carolina), Air Force Systems Command, and F-111 System Project Office (SPO), Aeronautical Systems Division.

In November 30, 1966, the planning, cost and schedule for a potential RDT&E and production program for the Navy RF-111B program was presented to Navy Rear Admiral Sweeney.

RF-111A prototype (F-111A No. 11 63-9776) was moved on January 30, 1967, from the factory to the special projects area to begin modification. It was completed in October 1967, and turned over to field operations at General Dynamics (GD), Fort Worth, Texas.

Also during October 1967, a firm price proposal was submitted for re-orientation of the RF-111A prototype program to an RF-111D configuration by incorporation of Mark II avionics, and a limited engineering go-head was authorized for RF-111 Group A provisions in production for the first six RAAF F-111Cs. By December, the efforts to obtain early funding or approval to initiate the RF-111D program were unsuccessful. The USAF plan for RF-111D sixty airplane procurement was still programmed to start with FY 1970 funding. In November 1968, GD, Fort Worth was directed to proceed with analysis to define the configuration and the task to modify an F-111D into an RF-111D.

This view of the one and only USAF RF-111A shows the weapons bay reconnaissance pallet. (USAF)

CHAPTER ONE: TFX PROGRAM - THE F-111

The reconnaissance pallet in the weapons bay of RAAF RF-111C A8-134 is visible in this photo. (82nd Wing Photographic, RAAF)

On December 17, 1967, the RF-111A prototype reconnaissance aircraft made a successful maiden flight. All sensors were operated and no problems were observed. The second flight of the aircraft was made on December 29, 1967. The RF-111A prototype, in the following six months, would make a series of subsonic flights over photographic and infrared resolution targets and radar reflectors located as James Connally AFB, Waco, Texas. The flights were conducted at altitudes as low as 1000 ft. and as high as 30,000 ft.

In March 1968, GD, Fort Worth division was directed by the SPO to reduce the flight test program on the RF-111A prototype from 12 months to 8 months so that the resultant credit could be reapplied to the initiation of the RF-111D program. In July 1968, the first flight of a planned test program at Eglin AFB, Florida was run against the underbrush range. Approximately one half of the planned runs were accomplished, the remainder were precluded by adverse weather. The testing continued through September 1968, when all RF-111 radar flights over the Eglin AFB underbrush range were completed. Excellent radar performance was achieved on the last two flights. This completed the first flight test portion of the RF-111A prototype program. On October 30, 1968, the RF-111 aircraft was flown to Shaw AFB where it was displayed to the personnel of the Tactical Air Reconnaissance Center. Upon returning to Fort Worth, the dismantling of the reconnaissance capability was initiated. With no further flying planned in the RF-111A prototype flight test program, the aircraft was stored in November 1968. It was pulled out of storage in March 1969, and activated for use in the flight test chase aircraft pool. The F-111 System Program Office (SPO) had planned to program additional development tests on this airplane in support of the side looking radar development for the RF-111D in response to Headquarters USAF request for a comprehensive analysis of the various coherent and non-coherent side looking radars which were being considered for installation in the RF-111D. The flyaway cost of the sole RF-111A was $12.1 million. Reconverting the RF-111A to its basic attack configuration proved impractical and the aircraft was retired following testing.

In September 1969, the Air Force, having little success in obtaining funds for a go-head on production of the RF-111, concluded that the plan to convert F-111A aircraft to the RF-111As, instead of buying new RF-111D or RF-111F aircraft, was a more supportable option.

The SPO, in October 1969, directed GD to submit a definition and proposal for converting F-111As to RF-111As. The conversion configuration of the RF-111A was to be essentially that which was tested during the RF-111A prototype program. All 159 F-111A test and production aircraft were to be assessed to determine the 46 F-111As which were the most logical to convert to the reconnaissance version. As a part of this study the 72 F-111As most cost effective to remain as fighters were to be identified. The modifications required to bring all RF aircraft to a common baseline and a study of design changes was to be included. The preliminary assessment of the F-111A fleet to determine the most cost effective 72 fighters and 46 reconnaissance aircraft was reviewed with F-111 SPO in November. As a result, on December 10, 1969, SPO direction was given to base the RF-111A proposal on converting the last 46 F-111As (67-0069 through 67-0114). At the same time they requested an alternate proposal for the last 24 F-111A aircraft (67-0091 through 67-0114). The RF-111 configuration definition and proposal preparation. The proposal (to convert 46 and an alternate 24 F-111A aircraft To RF-111A) was dropped in March 1970. As a result of the cancellation of the RF-111 models, the USAF continued to use its RF-4C's in the manned Tactical Reconnaissance role for the next 25 years. In July 1970, a sales program was presented based on substituting a small group of RF-111A aircraft for RF-4C aircraft instead of replacing RF-4Cs lost to attrition, but this too was not accepted.

The RAAF modified four of their F-111Cs to RF-111C configuration during 1978 through 1980, and continued to use the highly successful RF-111C.

TOTAL F-111 AIRCRAFT PRODUCTION
Total F-111 Aircraft Of All Types Produced – 562

F-111 Tactical Fighter Bombers (479)
(Not including F-111Gs)

Pre Production F-111A Test Aircraft
63-9766 through 63-9783 (18) aircraft

Production F-111A Test Aircraft
65-5701 through 65-5710 (10) aircraft

Operational F-111A Aircraft (131)
66-0011 through 66-0058 (48) aircraft
67-0032 through 67-0114 (83) aircraft

EF-111A Aircraft (42) (Modified from F-111As)
66-0013, 60-0014, 66-0015, 66-0016, 66-0018, 66-0019, 66-0020, 66-0021, 66-0023, 66-0027, 66-0028, 66-0030, 66-0031, 66-0033, 66-0035, 66-0036, 66-0037, 66-0038, 66-0039, 66-0041, 66-0044, 66-0046, 66-0047, 66-0048, 66-0049, 66-0050, 66-0051, 66-0055, 66-0056, 66-0057, 67-0032, 67-0033, 67-0034, 67-0035, 67-0037, 67-0038, 67-0039, 67-0041, 67-0042, 67-0044, 67-0048, 67-0052

F-111C Aircraft For Royal Australian Air Force (24)
67-0125 (A8-125) through 67-0148 (A8-148) (24) aircraft

F-111D Aircraft (96)
68-0085 through 68-0180 (96) aircraft

F-111E Aircraft (94)
67-0115 through 67-0124 (10) aircraft
68-0001 through 68-0084 (84) aircraft

F-111F Aircraft (106)
70-2362 through 70-2419 (58) aircraft
71-0883 through 71-0894 (12) aircraft
71-0895 through 71-0906 (12) aircraft canceled – Not built
72-1441 through 72-1452 (12) aircraft
73-0707 through 73-0718 (12) aircraft
74-0177 through 74-0188 (12) aircraft
75-0210 through 75-0221 (12) aircraft canceled – Not built

F-111G Aircraft (34) (Modified from FB-111As)
67-0162, 67-7193, 67-7194, 67-7196, 68-0241, 68-0244, 68-0247, 68-0252, 68-0254, 68-0255, 68-0257, 68-0259, 68-0260, 68-0264, 68-0265, 68-0270, 68-0271, 68-0272, 68-0273, 68-0274, 68-0276, 68-0277, 68-0278, 68-0281, 68-0282, 68-0289, 68-0291, 69-6503, 69-6504, 69-6506, 69-6508, 69-6510, 69-6512, 69-6514

In this view of the assembly line at Fort Worth, Texas, aircraft 160 (F-111E 67-0115) is visible in the foreground. (Lockheed Martin Tactical Aircraft Systems)

FB-111A Strategic Bombers (76)

Production Aircraft
67-0159 through 67-0163 (5) aircraft
67-7192 through 67-7196 (5) aircraft
68-0239 through 68-0292 (54) aircraft
69-6503 through 69-6514 (12) aircraft

F-111B Aircraft (7)

Pre Production F-111B Test Aircraft
151970 through 151974 (5) aircraft
152714 and 152715 (2) aircraft
152716 and 152717 (2) Not completed
153623 to 153642 (20) Not built – contract canceled
156971 to 156978 (8) Not built – contract canceled

F-111K (None Delivered)

F-111K
67-0149/XV884 The aircraft was not completed and was disassembled after the contract was canceled. The Crew module was used as part of the RF-111 program.

TF-111K
67-0150/XV885 The aircraft was not completed and was disassembled after the contract was canceled. The wing box was used in F-111E 68-0083, the forward and center fuselage was used as part of the RF-111 program, and the crew module was used as a trainer.

TF-111K Canceled/Construction not started.
67-0151 to 67-0152/XV886 to XV887

F-111K Canceled/Construction not started.
68-0152 to 68-0158/XV902 to XV947
68-0181 to 68-0210
68-0229 to 68-0238

THE GENERAL DYNAMICS, FORT WORTH F-111 RESTORATION FACILITY
(Aardvark Hospital)

General Dynamics, Fort Worth, Texas set up a repair/restoration facility for badly damaged F-111s. Parts from stored or scrapped aircraft were in the restoration. The following were rebuilt at the Forth Worth Facility:

F-111As	–	67-0079 and 67-0101
F-111Ds	–	68-0095, 68-0101, 68-0127, 68-0136, 68-0148, and 68-0174
F-111Es	–	67-0118 and 68-0082
FB-111As	–	67-7194 and 68-0259

The three photos on this page are of the rebuild of FB-111A 67-7194. 67-7194 sustained major damage as a result of hitting the approach lights and hard landing at Pease AFB on February 25, 1976. A fire also added to the damage. The aircraft was rebuilt at General Dynamics, Fort Worth using parts of other Aardvarks. The rebuild started on September 1, 1978, with front section from 67-7194 being mated to the aft section of 67-0160, along with the vertical stabilizer from FB-111A 69-6513, which was disassembled at Sacramento ALC, while going through Programmed Depot Maintenance (PDM). Because of the use of other F-111 parts in the rebuild, the 67-7194 was nicknamed Frankenvark. The aircraft returned to flying status with the 380th Bomb Wing in September 1980.

(Three photos - Chris McWilliams Collection)

CHAPTER TWO

The F-111 Design

The F-111 design incorporated four new technologies never before used in a production aircraft. General Dynamics Model number 12, which became the F-111, used the variable sweep wing and afterburning turbofan engines recommended by the Air Force Design requirements. The aircraft would also be the first to use a capsule ejection system, although other aircraft like the B-58 and XB-70 had encapsulating ejection seats, the F-111A was the first to have the full cockpit separate from the aircraft on ejection. The USAF and RAAF versions were the first production aircraft to use an all weather Terrain Following Radar (TFR) system.

Much of the F-111 variable sweep wing design technology evolved from Bell's potbellied X-5, America's first swingwing airplane, and the US Navy's Grumman XF10F. The F-111's two-pivot, variable-sweep wing, as opposed to the single-pivot used in previous experiments, was the successful design of the variable-wing idea, and later used in the F-14, B-1, and the Panavia Tornado.

Experiments changing the flight characteristics of an aircraft by altering its wing sweep while airborne had been tried many times before the F-111 design. The first airplane to use variable wing-sweep was tested in 1931. It was the Westland-Hill Pterodactyl IV, a tailless monoplane whose wing could be moved almost five degrees during flight. During the Second World War, the Germans also investigated the variable geometry principle. The Messerschmitt P 1101 prototype design work began in July 1942. The aircraft was approximately 80% complete when captured by the Allied forces in April 1945. The Bell Aircraft X-5 was based on this German swept-wing fighter prototype. The captured P 1101 aircraft was shipped to Wright Field near Dayton, Ohio. The aircraft was modified to test the movable-wing principle. On June 20, 1951, the Bell X-5 flew for the first time.

The Navy felt the variable-geometry wing was a natural for carrier operations. With the wings forward, the takeoff and landing characteristics would be ideal for carrier operations, and the swept-back position of the wings would permit high speed flight. The first Navy plane to use variable sweep wings was the Grumman's XF10F-1 Jaguar. Two prototype aircraft were assembled and one flown in the test program. The first flight occurred on May 19, 1952. The program did not meet expectations, and a planned buy of 135 aircraft was canceled. Even though the Jaguar program was canceled, the most advanced feature of the XF10F-1, the variable-sweep wing, functioned satisfactorily. The results of the Jaguar program were highly valuable and used by Grumman in design of the F-111B.

The Pratt & Whitney TF30 was the first production afterburning turbofan engine. Earlier afterburning jet engines ducted the engine exhaust through an afterburner (basically a ram jet engine) where more fuel was added and burned providing additional thrust. The turbofan jet engine provides thrust from two sources; the engine exhaust, as in a turbojet; and the air ducted from the forward fan stages of the engine. The TF30 channels both the engine exhaust and the ducted fan air into the afterburner, providing more efficient power than that available from an afterburning turbojet.

Due to the planned high supersonic speeds and the requirement for high subsonic speed in the low altitude-high dynamic pressure (high Q) flight regime, a standard zero-zero ejection seat system like those used in the other Century Series Aircraft was inadequate. An ejectable crew module was designed into the fuselage. This module, or capsule as it was also called, housed both crew members, and had both a zero-zero and supersonic ejection capability.

The Terrain Following Radar (TFR) system was a radar system which detected the terrain in front of the aircraft. It provided guidance to the crew, and input to the autopilot, which allowed the crew to fly the aircraft safely at very low level, as low as 200 feet above the ground, and through the whole speed regime of the aircraft. Differing from the terrain avoidance/terrain following systems of other aircraft, including the B-52, the RF-4C, and the B-1B, the F-111 TFR is a stand alone system, with its own antennas, and is not part of the aircraft's attack radar system.

CHAPTER TWO: THE F-111 DESIGN

DESCRIPTION

The F-111 series aircraft are two place (side-by-side) long range fighter bombers built by General Dynamics, Fort Worth Division. The aircraft were designed for all-weather supersonic operation at both low and high altitude. Mission capabilities of the Tactical Fighter Bomber included: long range high altitude intercepts utilizing AIM-9 air-to-air missiles and/or M61A1 Vulcan gun (if installed); long range attack missions utilizing conventional or nuclear weapons as primary armament (the SAC FB-111As could also carry and launch Boeing AGM-69A Nuclear Short Range Attack Missiles - SRAMs); and close support missions utilizing a choice of missiles, free fall and guided bombs, and rockets. The U.S. Navy F-111B with its AN/AWG-9 fire control system, primarily designed as a fleet defense aircraft, carried AIM-54/A Phoenix missiles. The RAAF F-111Cs have been modified to carry AGM-84 Harpoon anti-shipping missiles and AGM-142 air to ground optically guided missiles.

ENGINES

The F-111s are powered by two Pratt & Whitney TF30 sixteen-stage axial flow turbofan engines equipped with afterburners. The engines are mounted side by side in the fuselage and are interchangeable.

The engines were designed for starting using an external pneumatic ground starter cart. Also the left engine has the capability of being started by means of a pyrotechnic cartridge. With either engine operating, the other engine can be started by using bleed air from the operating engine. Electrical power is supplied for the engine igniter plugs by an engine driven alternator. Each engine is supplied air through inlet duct located below the intersection of the wing and fuselage on the same side of the aircraft as the engine.

Inlet System

Engine inlet air flow is regulated throughout the entire aircraft speed range in order to maintain maximum engine performance. This regulation of the air flow is accomplished

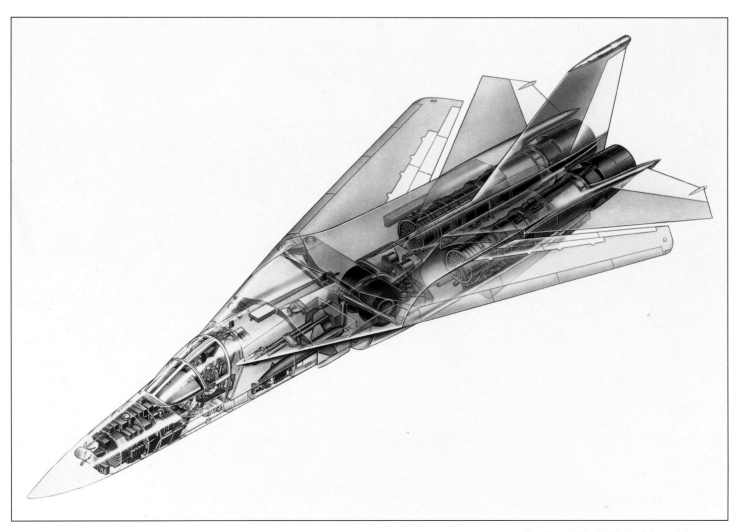

(Lockheed Martin Tactical Aircraft Systems)

The Engine

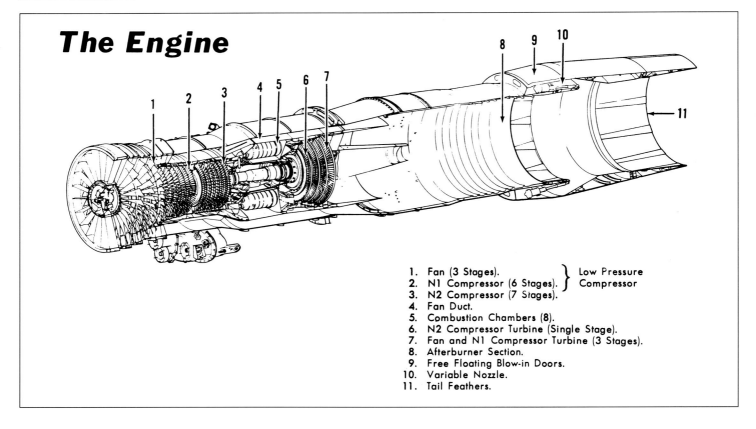

1. Fan (3 Stages). } Low Pressure
2. N1 Compressor (6 Stages). } Compressor
3. N2 Compressor (7 Stages).
4. Fan Duct.
5. Combustion Chambers (8).
6. N2 Compressor Turbine (Single Stage).
7. Fan and N1 Compressor Turbine (3 Stages).
8. Afterburner Section.
9. Free Floating Blow-in Doors.
10. Variable Nozzle.
11. Tail Feathers.

by a movable spike located in the inlet of each engine. Each spike is a quarter circle, conical shaped, variable diameter body that is independently movable forward and aft. The spikes are located in each air intake at the intersection of the wing lower surface and the fuselage boundary plate. Position and shape of the spikes are changed automatically to vary the inlet geometry and to control the inlet shock wave system. Local air pressure changes due to variations in Mach pressure ratios are sensed by the spike control unit. Signals from the control unit operate hydraulic actuators which are powered by the utility hydraulic system to position the spike fore and aft (extend or retract) and adjust the spike cone angle by contracting and expanding the spike as required. In the event the system malfunctions, a one-shot pneumatic override system is provided to position and lock the spike full forward and fully contracted.

During ground operation and low speed flight, an additional amount of air is required to prevent possible compressor stalls. This air is provided by translating cowls or cowl blow-in doors. The F-111A/EF-111A, the first FB-111A (67-0159), the first five F-111Bs (151970 through 151974), and the F-111C provided the additional air through the use of translating cowls. The cowls slide forward creating openings in the outboard side of each nacelle allowing the additional air

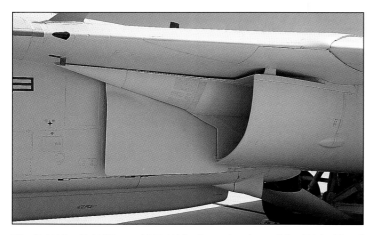

The F-111As, F-111Cs, and EF-111As (seen here) used the Triple Plow I intake system. (Author)

The Triple Plow II intake system used on F-111Ds, F-111Es, F-111Fs, and FB-111As is seen on this F-111F. (Author)

to reach the engine. The cowls are actuated using hydraulic pressure from the utility hydraulic system. The cowls can be opened pneumatically in the event of a hydraulic system failure. During normal operation the cowls are automatically controlled as a function of speed and altitude. They can also be opened manually.

The engine inlet cowls on the F-111B (152714 and 152715), F-111D, F-111E, F-111F, F-111G, and FB-111As (67-0161 and on) are equipped with three blow-in doors on each engine cowl to provide an opening for additional air to the engine when needed. FB-111A 67-0160 (the second FB-111A) was equipped with an experimental intake called "Superplow" which had only two blow-in doors. The doors are free floating, and as a result, automatically assume a position corresponding to the pressure differential between inlet duct and outside air pressure. On the ground and during low speed flight, the doors will move to a position between full open and closed as required by engine demand. As speed is increased the doors will close and remain closed as long as ram air flow keeps the inlet pressure above the outside air pressure.

Splitter plates, used with translating cowl equipped aircraft, and boundary layer diverter ducts used with blow- in door-equipped aircraft at the front of the inlet ducts remove the low energy air from the fuselage and the lower surface of the wing glove, preventing boundary layer air from disturbing engine inlet air. The combination of the translating cowl and the splitter plates is referred to as the Triple Plow I intake system; blow in doors with boundary layer diverter ducts are referred to as the Triple Plow II intake system.

These features allow optimum engine performance throughout a wide range of aircraft operating conditions. Air from the inlet of each engine is routed through a single duct for both the basic engine section and the fan section. Three compressor stages provide the initial pressurization of the air flowing into the engine and into the fan duct. The fan duct is a full-annular duct which directs air aft to join the engine airflow coming from the turbine discharge. The fan air develops a significant portion of the total engine thrust. Engine air is compressed by 9 stages of the low pressure compressor (N1) of which 3 stages are the fan, and 7 stages of the high pressure compressor (N2). The air is then diffused into the combustion section which contains 8 combustion chambers. The turbine section of the engine consists of a single stage turbine to drive the high pressure compressor and a three-stage turbine to drive the low pressure compressor. The turbines are mechanically independent of each other. High pressure compressor speed is indicated by a tachometer. Speed of the low pressure compress is not monitored except by an overspeed caution lamp. After leaving the turbine section of the engine, the air is joined with the fan air in the afterburner section. Engine compressor bleed air from the sixteenth stage is used for pressurization, air conditioning, and engine inlet anti-ice functions. Twelfth stage compressor bleed air is used for engine inlet guide vane anti-icing. Seventh stage compressor bleeds open under certain conditions to prevent compressor stall.

The ingestion of foreign objects into the engine was prevented by an aerodynamic screen of engine bleed air, which is directed down and outboard beneath each inlet through vortex destroyer air jets. The vortex destroyers serve to prevent the formation of vortexes below the inlet, thereby preventing foreign objects from being captured in a vortex and sucked into the engine. When the weight of the airplane is on the landing gear, a ground safety switch located on the landing gear automatically activates the vortex prevention air screen.

Afterburner

The afterburner (AB) augments thrust by injecting fuel into the engine exhaust stream in the afterburner section where it is ignited by a hot streak ignition system. This ignition system injects extra fuel into combustion chamber 4. The fuel is ignited by the combustion chamber fire. The extra fuel causes a longer flame which burns past the turbine section and lights the afterburner. The throttle position, through the afterburner fuel control, varies the amount of fuel supplied to the afterburner.

Engine Types

The F-111A/EF-111A, F-111C, and F-111E were powered by two TF30-P-3 engines with a sea level, standard day static thrust rating of 10,750 pounds in military power and 18,500 pounds in afterburner. The F-111B was powered by two TF30-P-12 engines with a sea level, standard day static thrust rating of 12,290 pounds in military power and 20,250 pounds in afterburner. The F-111D was powered by two TF30-P-9 with a sea level, standard day static thrust rating of 12,000 pounds in military power and 19,600 pounds in afterburner. The F-111F was powered by two TF30-P-100 with a sea level, standard day static thrust rating of 14,560 pounds in military power and 25,100 pounds in afterburner. The FB-111A was powered by two TF30-P-7 with a sea level, standard day static thrust rating of 12,290 pounds in military power and 20,250 pounds in afterburner.

The P-3 engines in the F-111A, EF-111A, F-111C, and F-111E were upgraded to the P-109 configuration, as were the P-9 of the F-111D and P-100 of the F-111F. The EF-111As and the F-111Es received the P-109 configuration while assigned to the 27th TFW at Cannon AFB, giving all aircraft of the wing the same type engine. The P-7 engines of the FB-111A were upgraded to P-107 configuration. RAAF F-111Gs are undergoing re-engining from the P-107 to a RAAF hybrid engine unofficially referred to as the P-108. The P-108 has the forward section of the P-109 and the aft section of the original P-107.

AIRFRAME

The airframe is a semi-monocoque structure composed of upper and lower longerons, frames, bulkheads, and vertical stiffeners. Except for the nose radome, which is glass fabric, the fuselage is covered with sandwich skin panels, two-layer skin panels, and single-layer skin panels composed of aluminum alloys, corrosion resistant steel, titanium alloys, plastics, and resins, as required to reduce thermal buckling and deformation. The variable sweep wings are mounted at the top of the fuselage above the engine intakes. The horizontal stabilizers are capable of symmetrical or differential movement. The F-111 uses a single vertical stabilizer and rudder.

Wings

The wings are unique in design and mounting. Two different wing sets were used on the F-111s. The F-111A, F-111D, F-111E, and F-111F had "short" wings with a wing span, at 16 degree sweep of 63.0 feet. The F-111B, F-111C, and FB-111A "long" wing aircraft had a wing tip extension added lengthening the wings to a span of 70.0 feet at 16 degree sweep. Ferry wing tip extensions had been designed and were available for the short winged F-111, but were rarely used.

Each wing is full cantilever, variable sweep in design, and incorporates spoilers, rotating gloves, flaps and slats. Primary structure consists of a wing box and pivot fitting. The wing box is composed of multiple spars, bulkheads and skins and provides mounting for wing-mounted flight control surfaces. The internal box area forms the wing fuel storage tank. Surfaces which are attached to the wing box include the removable short or long wing tip, variable camber leading edge slats, spoilers, rotating glove, and trailing edge flaps. The pivot fitting provides mounting to the fuselage and permits wing sweep movement from 16 to 72.5 degrees. Four removable external stores pylons can be carried on the underside of each wing. The inboard two pylons on each wing swept with the wing. The outboard two were fixed requiring the wing sweep to be locked when they were carried.

Each wing is equipped with full span, five-section Fowler-type flaps. The outer four sections, designated as the main flaps, are mechanically interconnected and operate as one unit. The inboard flap, designated as the auxiliary flap, operated independently from the main flaps, and was later deactivated. The main flaps are powered by a hydraulic motor during normal operations and an electric motor during emergency operations. The auxiliary flaps were independently operated by electric actuators. The auxiliary flap actuators were disabled whenever the wing sweep angle is greater than 16 degrees. Asymmetrical flap travel is prevented by an asymmetry device.

Each wing is equipped with four air deflector doors (five on long wing models) which are mounted on the lower surface of the wing just forward of the flaps. The doors are mechanically connected to the flap linkage and function as seals when the flaps are up and as air deflectors when the flaps are down.

Each wing is equipped with four-section leading edge slats (five on long wing models) which are mechanically interconnected and operate as one unit. The slats operate in conjunction with the main flaps. Asymmetrical slat travel is prevented by an asymmetry device.

Each wing is equipped with inboard and outboard hydraulic-operated spoilers. The inboard spoilers are operated by the primary hydraulic system and the outboard spoilers are operated by the utility hydraulic system. The emergency systems of each actuator are supplied by the opposite hydraulic system. The spoilers serve as flight controls and ground roll spoilers. As flight controls, the spoilers are used in conjunction with the horizontal stabilizers to control the aircraft around the roll axis when the wings are in a sweep angle of less than 45 degrees. The spoilers are deactivated when the wings are swept more than 45 degrees.

The outboard edges of the wing gloves adjacent to the wing inboard leading edges are equipped with movable surfaces to allow full forward movement of the inboard slats. These surfaces are called rotating gloves. A door forms the lower surface of each rotating glove. Each rotating glove and its associated door are operated by a mechanical actuator and linkage which are connected to the slat drive flexible shaft. When the slats are extended, the rotating gloves automatically rotate (leading edge down and trailing edge up) and the doors open to allow full extension of the slats.

Wing Sweep System

The variable sweep wings are moved and held in position by two hydraulic, motor-driven, linear actuators attached to each wing. The actuators are mechanically interconnected to ensure positive synchronization. The right actuator on each wing is furnished power by the utility hydraulic system, and the left actuator on each wing is furnished power by the primary hydraulic system. In the event of failure of either hydraulic system, the remaining system, by utilizing the load transfer capability of the mechanical interconnect, will still provide wing actuation. However, actuation under this condition will be at a reduced rate commensurate with actuator loading. Wing position is controlled by a closed loop mechanical servo-system in response to an input signal from the wing sweep handle. The wing sweep handle is locked in the 16 degree position whenever the auxiliary flaps are out of the zero position. A mechanical interlock prevents the wing sweep handle from being moved past the 26 1/2 degree position when either the flap or slat handle is out of the UP position or the main flaps are out of the fully retracted position. A 16 degree lock is also installed. This lock was intended to prevent sweeping the wings aft of 16 degrees when the auxiliary flaps, which were later deactivated, were out of the zero position. A loss of power to the solenoid which operates the lock will result in locking the wing sweep handle at 16 degrees, if moved to this position.

Vertical and Horizontal Stabilizers

The vertical stabilizer consists of a fixed fin, utilizing a primary box structure composed of multiple spars and aluminum skins, with a fixed leading edge and equipment carrying

CHAPTER TWO: THE F-111 DESIGN

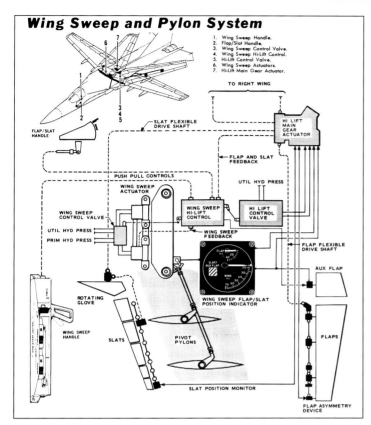

Wing Sweep and Pylon System

1. Wing Sweep Handle.
2. Flap/Slat Handle.
3. Wing Sweep Control Valve.
4. Wing Sweep Hi-Lift Control.
5. Hi-Lift Control Valve.
6. Wing Sweep Actuators.
7. Hi-Lift Main Gear Actuator.

tip. Rudder hinge supports are attached to the rear spar. The stabilizer also functions as an integral fuel system vent tank. There are no fixed horizontal stabilizers as such, since the entire horizontal surfaces (stabilizers) pivot. When both horizontal surfaces move together in the same direction, they act as elevators; and when they move in opposite directions, they act as ailerons. The stabilizers are attached to fuselage pivot fittings. Symmetrical movement provides pitch control while asymmetrical movement provides roll control.

Overwing Fairing and Pneumatic Seals

The overwing fairing and wing-fuselage pneumatic seals close the void in the fuselage above, below, and aft of the wing at all angles of wing sweep. When inflated, the pneumatic seals prevent airflow into or out of the fuselage cavity. The seals receive pneumatic pressure through a pressure regulator at all times when either engine is operating or when ground servo air is supplied to the aircraft.

Strakes

Fixed strakes are located on the lower portion of the engine access doors at approximately 30 degrees from vertical. They augment directional stability at supersonic speeds and high angle-of-attack operations.

Nose Radome

The nose section of the aircraft consists of a nose radome which can opened. The radome houses the attack radar and TFR antennas in all but the F-111B, and the Phoenix Missile System antennas in the F-111B. The radome is constructed of high temperature, resin-bonded, fiber-glass material and has an anti-static, thermally reflective, rain erosion resistant coating.

FLIGHT CONTROL SYSTEMS

The flight control systems are comprised of the primary controls, the flaps and slats, and the wing sweep subsystems. The primary flight controls consist of the movable horizontal tail surface, which controls pitch and roll movement; four spoilers which also control roll movement and, when the aircraft is on the ground, reduce landing roll; and the rudder, which controls aircraft yaw. The flap and slat sub-system consists of wing trailing edge flaps and wing leading edge slats, which provide additional wing area for takeoffs and landings. The wing sweep sub-system consists of actuators and mechanical linkage and controls to provide wing sweep movement from an angle of 16 to 72.5 degrees. This provides the aircraft with a wide range of flight capabilities.

LANDING GEAR SYSTEMS

The F-111 has a tricycle-type retractable landing gear, and an operational tail bumper and arresting hook. Two types of landing gear systems were used on the F-111 series. They appeared identical, and functioned in the same manner, but were capable of supporting different maximum takeoff weights. The light weight gear was used on the F-111A, F-111D, F-111E, and F-111F and limited the takeoff weight to approximately 85,000 pounds. The F-111B. F-111C, and FB-111A heavy weight gear allowed the takeoff weight to be increased to 114,000 pounds.

The landing gear systems are powered by the utility hydraulic power and 28 volt dc electrical power systems. Pneumatic pressure supplies the power for emergency extension and locking of the main and nose landing gear and emergency lowering of the tail bumper. The main landing gear is a single unit with a left and right wheel and tire assembly equipped with multiple-disk-type hydraulic brakes incorporating an anti-skid system. A main landing gear forward door/speedbrake operates during normal extension and retraction of the landing gear and independently as a speed brake during flight with the landing gear in the retracted position. The nose landing gear has dual tires, and is equipped with a hydraulic nose wheel steering unit.

The tail bumper extends and retracts in conjunction with the landing gear and protects the control surfaces, engines, and aft portions of the airframe from damage during aircraft ground operations.

ARRESTING HOOK

The arresting hook system is used for emergency landings. Except for the controls, the arresting hook components are located in the lower aft end of the fuselage tail cone. The hook is forced down by gravity and dash pot (accumulator) pressure. The hook on all models except the F-111B had to be manually stowed on the ground after landing.

FUEL AND IN-FLIGHT REFUELING SYSTEMS

The fuel system consists of integral fuselage and wing tanks, provisions for externally mounted tanks, and the necessary pumps, plumbing, valves, and control devices for fuel transfer and engine supply under all conditions. A trap tank is also provided as an integral part of the forward fuselage fuel tank and incorporates provisions for automatic refilling and engine suction feed. All the fuel tanks are vented together and into the vertical fin box section which is utilized as a vent tank and/or fuel overflow tank. The fuel system is equipped with an electrically controlled, fuel pressure-operated dump system. A unique feature of the fuel system is the ground check provisions which are provided near the single point refueling receptacle for determining that the high level shutoff valves are operating properly prior to ground refueling.

In-flight refueling in the USAF and RAAF F-111s is accomplished from a flying boom type tanker through a receptacle which opens on the top of the fuselage behind the pilot. The F-111B and F-111K designs were fitted with a retractable probe for probe and drogue refueling. The probe deployed out of the fuselage ahead of the windshield.

All the F-111 models except the Navy F-111B and British F-111K used the boom and receptacle inflight refueling system seen here. (Author)

CHAPTER TWO: THE F-111 DESIGN

Fuel Quantity and Tank Arrangement (*Typical*)

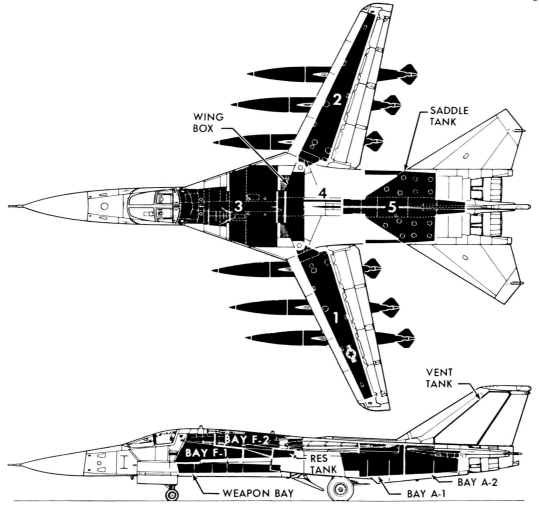

	LOCATION	QUANTITY			
		USABLE FUEL		FULLY SERVICED	
		GALLONS	POUNDS	GALLONS	POUNDS
1	LEFT WING INTERNAL TANK	389.2	2,530	390.7	2,540
2	RIGHT WING INTERNAL TANK	389.2	2,530	390.7	2,540
3	FORWARD FUSELAGE TANK	2,808.3	18,254	2,822.9	18,349
4	FUEL LINES	28.3	184	53.4	347
5	AFT FUSELAGE TANK	1,428.8	9,287	1,430.9	9,301
	TOTAL	5,043.8	32,785	5,088.8	33,077

NOTES:
1. These are average figures based on single point refueling at normal ramp altitude. Weights based on JP-4 fuel at 6.5 pounds per gallon. (Std. Day Only).

2. Each external tank, when carried, will have the following capacities.

USABLE FUEL		FULLY SERVICED	
GALLONS	POUNDS	GALLONS	POUNDS
PIVOT PYLONS			
601.2	3,908	603.4	3,922
FIXED PYLONS			
603.2	3,921	605.4	3,935

3. Weapons bay tanks, when carried, will have the following capacities.

USABLE FUEL		FULLY SERVICED	
GALLONS	POUNDS	GALLONS	POUNDS
LEFT WEAPONS BAY TANKS			
269.2	1,750	272.3	1,770
RIGHT WEAPONS BAY TANKS			
290	1,885	293.1	1,905

AE000000-F015

CREW MODULE SYSTEM

General Dynamics believed the F-111's crew module (first known as "boiler plate" crew escape capsule) ranked alongside the F-111's variable-sweep wing and afterburning fan engine as major advancements in aircraft design. Developed by the McDonnell Aircraft Corporation and initially tested in February 1966, the F-111's crew module was fully automated. All production F-111 aircraft, employed the ejectable crew module. The crew module forms an integrated portion of the forward fuselage, the crew compartment and the forward portion of the wing glove. The system affords maximum protection for the crew members throughout the aircraft performance envelope and includes capabilities for safe ejections at maximum speed and altitude as well as at zero altitude and zero speed. Survival equipment and a recovery parachute are included as a part of the crew module system.

When forced to abandon his aircraft, the pilot or WSO only had to "squeeze, and pull" one lever. This caused an explosive cutting cord to shear the module from the fuselage; a rocket motor fired, ejecting the module carrying both crew members upward, and then was lowered by parachutes to the ground or sea. Upon land or water impact of crew module, landing shock reduction is provided by air bags attached to the module. The crew module is also provided with underwater escape capabilities, self-righting buoyancy, flotation capacity, and occupant environmental hazard protection. An emergency oxygen supply and emergency cabin pressurization supply are furnished as a portion of the crew module system.

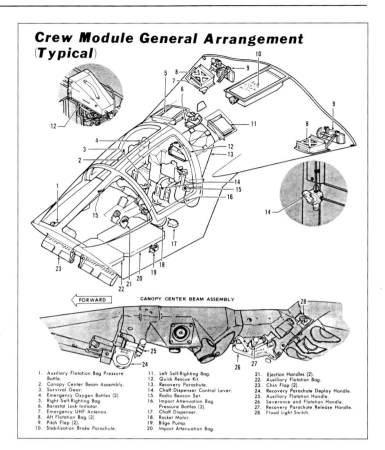

Crew Module General Arrangement (Typical)

1. Auxiliary Flotation Bag Pressure Bottle.
2. Canopy Center Beam Assembly.
3. Survival Gear.
4. Emergency Oxygen Bottles (2).
5. Right Self-Righting Bag.
6. Barostat Lock Initiator.
7. Emergency UHF Antenna.
8. Aft Flotation Bag (2).
9. Pitch Flap (2).
10. Stabilization Brake Parachute.
11. Left Self-Righting Bag.
12. Quick Rescue Kit.
13. Recovery Parachute.
14. Chaff Dispenser Control Lever.
15. Radio Beacon Set.
16. Impact Attenuation Bag Pressure Bottles (2).
17. Chaff Dispenser.
18. Rocket Motor.
19. Bilge Pump.
20. Impact Attenuation Bag.
21. Ejection Handles (2).
22. Auxiliary Flotation Bag.
23. Chin Flap (2).
24. Recovery Parachute Deploy Handle.
25. Auxiliary Flotation Handle.
26. Severance and Flotation Handle.
27. Recovery Parachute Release Handle.
28. Flood Light Switch.

The capsule from 67-0080, which was used successfully in an ejection on March 11, 1976, is seen here at Nellis in July 1976. The other capsule air bags, not used in the ejection, were tested. The blue air bags on the bottom and aft of the capsule are flotation bags used to keep the capsule afloat after a water landing. The yellow bags on the top of the capsule are the self righting bags used to ensure the capsule floats upright. (Author)

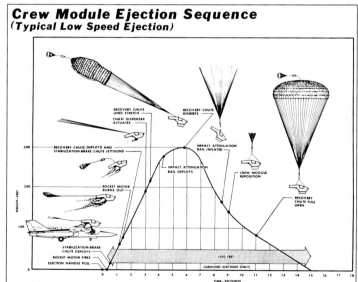

Crew Module Ejection Sequence (Typical Low Speed Ejection)

CHAPTER TWO: THE F-111 DESIGN

EJECTION SEATS

With the crew module not available at the beginning of production, ejection seats for both crew members were installed in the first 11 F-111A aircraft (63-9766 through 63-9776) and the first three F-111Bs (151970, 151971, and 151972). The seat was a rocket catapult ejection seat. The ballistically initiated rocket catapult operated independently of other airplane systems to permit emergency escape during flight. The seat was electrically adjustable vertically and horizontally. The seat headrest was a recessed V-type covered with foam rubber. A safety latch handle was located in the center of the headrest. It rotated downward to safety the seat ejection controls while on the ground. The latch had to be stowed to the flush (armed position) before the crew members head would fit in the headrest. The seat incorporated personnel restraint equipment composed of an inertia reel, shoulder harness and lap belt, a parachute with an automatic barometric time delay actuator, a survival kit, a seat pan cushion containing an emergency oxygen system, two pneumatic seat man separators, and two sets of ejection handles.

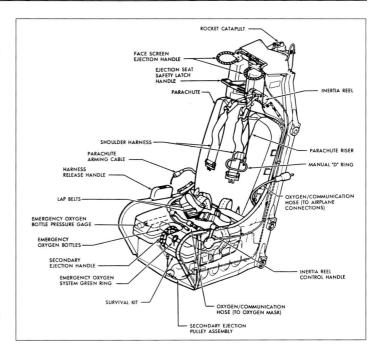

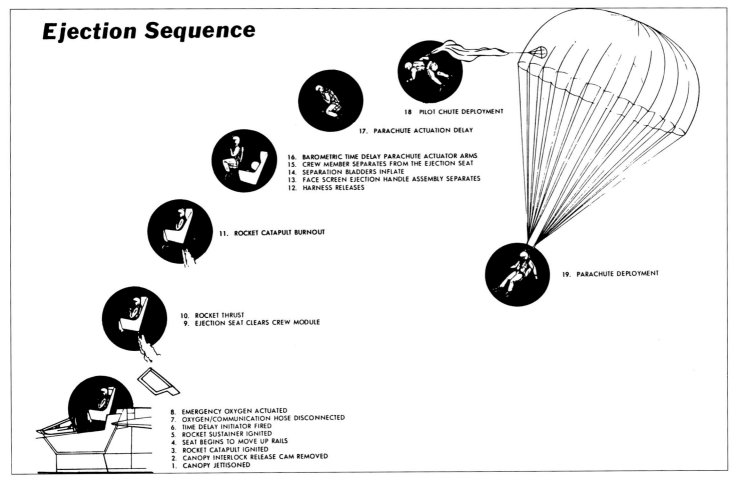

Ejection was accomplished by either pulling the face curtain rings down or by pulling the secondary ejection handle up. The first few inches of travel jettisoned the center beam and the full canopy for both crew members. As the canopy left the aircraft, a canopy/ejection handle interlock release cam was actuated allowing additional travel of the rings or handles to release the rocket catapult firing pin and eject both seats simultaneously.

If the canopy did not jettison with the ejection handles, the seats could not be ejected until the canopy was jettisoned with the canopy jettison T-handle, located on the canopy center beam.

During ejection the crew member was held firmly in the seat by the inertia reel shoulder harness which attached to the crew member's torso harness across the shoulders, and by the lap belt which attached to the torso harness on each side. As the seat started to move up the seat rails, the oxygen and communications lines were pulled free of the airplane systems and the emergency oxygen supply located under the seat pan was activated by a lanyard. After the seat left the aircraft, the restraints were released, and the seatman separator was actuated to separate the crew member from the seat. As the crew member separated from the seat, the automatic parachute time delay actuator was armed by a lanyard attached to the seat. The parachute automatically opened when below the preset barometric altitude. If the automatic parachute opener malfunctioned, and the crew member must pull the manual D-ring on the left shoulder strap of the parachute harness to deploy the parachute.

EGRESS/INGRESS STEP and ENTRANCE LADDER

A step and entrance ladder is located on each side of the fuselage below the cockpit area. The step and ladder can be released electrically from the cockpit. The ladder and step also can be released manually and must be retracted and locked manually. This system was installed on F-111As, F-111Cs, F-111Es and FB-111As. It rarely used and was deactivated on most aircraft.

AIR CONDITIONING AND PRESSURIZATION SYSTEMS

Two different installations were used in all F-111 models. The F-111A, F-111C, F-111E, and FB-111A having one type, with the F-111D, F-111F, and EF-111A having a different type. Both system types used engine bleed air and fulfilled the following functions:

1) Provide and control a supply of air for ventilation, pressurization, and temperature control of the cabin, pressure suits and anti-g suits.
2) Provide and control a supply of air for windshield clearing, and electronic equipment cooling and pressurization.

3) Provide a supply of air for engine inlet anti-icing, the foreign object damage (FOD) air curtain to the engines, aircraft fuel tank and wing seal pressurization, and pressurization of the hydraulic system and to the ejector for ground cooling of the hydraulic fluid.
4) Provide the ducting and fittings for pneumatic engine starting.
5) Supply emergency ram air.

HYDRAULIC AND PNEUMATIC POWER SYSTEMS

The F-111 hydraulic and pneumatic power systems provide hydraulic and pneumatic power for the various aircraft control systems and to the landing gear system. The hydraulic power systems supply power to hydraulic systems for normal aircraft operation. The electrical emergency generator is driven by a hydraulic power system. The pneumatic power systems provide emergency power for alternate landing gear extension. Pneumatic pressure for hydraulic reservoir pressurization is supplied by a pneumatic system. The hydraulic power system is composed of the primary and utility hydraulic power systems, hydraulic pressure indicating system, and hydraulic caution system.

The primary and utility hydraulic power systems are two independent and parallel systems. These are closed center, Type II, Class 3000 psi systems. Each system is equipped with the two engine-driven variable delivery hydraulic pumps to generate and supply operating pressure and flow to hydraulic subsystems and components in accordance with demand.

ELECTRICAL POWER AND LIGHTING SYSTEMS

The electrical power and lighting system for the F-111 aircraft is comprised of a main ac power system (supply and distribution), dc power system (supply and distribution), emergency ac power system, and interior and exterior lighting systems.

Primary ac power is supplied by two main ac generators. Each generator supplies constant 115/200-volt, three-phase, 400 hertz (normal) electrical power to the aircraft. The dc power is derived from the left main and essential ac buses by a main and essential ac to dc 150 ampere converter units. The 28 volt dc output of the converters is connected in parallel during normal system operation to supply power to the main and essential dc buses. An emergency ac generator, mechanically coupled to a hydraulically operated motor, automatically supplies 115/200-volt, three-phase, 400 hertz (nominal) ac power to the essential ac buses, for subsequent distribution to essential safety of flight loads in the event of a multiple failure of the main ac power system. In the emergency mode of operation, 28 volt power is only available from the essential converter unit for safety of flight loads. Operation of the emergency generator is terminated when either or both main generators are placed back on the line. When the aircraft is on the deck, external electrical power can be connected to supply ac power to the buses that are normally supplied by the main ac power supply system.

The exterior and interior lighting systems furnish the required illumination for night, adverse weather, formation, and high altitude operating conditions. The exterior lighting system consists of wing and tail position lights, formation lights, anti-collision lights, landing/taxi light, and approach lights. The interior lighting system (primary and secondary) consists of internally mounted and edge lights that provide the required primary illumination for pilot's and WSO's instrument panels, center, side, and overhead consoles. The secondary lighting system consists of floodlights and utility lights. The intensity of both primary and secondary lighting is controlled from the cockpit to satisfy operational requirements. A warning and caution system is provided to advise the crew of system/subsystem malfunctions.

AIR DATA COMPUTER SYSTEM

The all data computer system computes and transmits data concerning the immediate atmosphere through which the aircraft is traveling. These computations are performed by the Central Air Data Computer (CADC) and the Maximum Safe Mach Assembly (MSMA). The immediate atmosphere of the aircraft is sensed by the pitot-static probes, the total temperature probe, and the angle-of-attack transmitter. The pneumatic and electrical inputs are applied to the analog-computing systems (CADC and MSMA) which compute the aircraft flight parameters and apply them to the various aircraft systems and subsystems. A malfunction in the air data computer systems activates a CADS malfunction advisory light on the caution annunciator panel on the pilot's instrument panel.

AUTOMATIC FLIGHT CONTROL SYSTEM

The Automatic Flight Control System (AFCS) consists of rate gyros, accelerometers, flight control computers, and damper servos which, in conjunction with the primary flight control system, provide damping about the pitch, roll and yaw axes during flight. It also controls the pitch and roll axes during the various modes of autopilot operation. The automatic flight control system computers receive various input signals and compute command signals to the pitch, roll, and yaw dampers to control the aircraft. The control circuits are designed so that stability augmentation is engaged during all normal flight conditions. Incompatible mode selection is prevented by circuit interlocks. Attitude stabilization, the basic autopilot mode, is in effect any time an autopilot mode is engaged. The other modes, when engaged, merely modify the basic mode references to accomplish specific tasks. Attitude stabilization will hold the aircraft at the reference roll and/or pitch attitude until selection of another autopilot mode or until

the pilot initiates control stick steering. The pilot can manually maneuver the aircraft at any time by use of control stick steering without disengaging the autopilot switches.

BOMBING-NAVIGATION SYSTEMS

All F-111s shared similar air-to-ground radios, intercommunication systems, navigational radios, instrument lending systems, and central air data computers. They also had like identification equipment, flight control, and radar altimeter subsystems, as well as extensive electronic countermeasure and penetration aid equipment. Remaining avionics were quite different. For instance, the Mark I system (consisting of attack radar, navigation-attack system, and a lead computing optical sight), common to the F-111A, C, and E models, could not be compared to the Mark II that was being developed.

MK I AVIONICS SYSTEM F-111A, F-111E, and F-111E (AN/AJQ-20A)

The analogue Bombing-Navigation (Bomb-Nav) system is a self contained dead reckoning inertial system. The system consists of a Stabilized Platform, a Navigation Computer, a remotely located flux valve, a Bomb Nav Distance Time Indicator (BNDTI), and an automatic Ballistics Computer Unit (BCU). A removable offset panel was developed as a result of aircrew experience in combat missions in Southeast Asia. the panel can be installed in the right main instrument panel to provide a multiple, pre-programmed offset aimpoint capability. Pre-programming is accomplished by removing the panel prior to flight and setting the potentiometers in the panel.

The Stabilized Platform (SP), located in the forward electronics bay, consists of a four-gimbal, all attitude inertially stabilized platform, and its associated electronics. The SP supplies outputs of pitch, roll, true heading, vertical acceleration, and north and east components of ground speed. Additionally, signals are provided to indicate (1) progress of initial alignment, (2) proper range of SP gyroscope temperatures, and (3) reliability of SP output data. Prior to flight, the SP is initially aligned.

The Navigation Computer (NC), located in the right main instrument panel, is a self-contained navigation steering, radar sighting and bombing computer. The primary inputs to the NC are north and east components of ground speed, true heading and pitch angle from the stabilized platform. True air speed and pressure altitude from the CADC and magnetic heading from the system flux valve. The NC provides all the computing, control, and display functions for the Bomb Nav system. In event of SP failure, the NC will continue to operate using last computed wind or manually entered wind combined with true airspeed from the CADC and magnetic heading from the Auxiliary Flight Reference System (ATRS) to substitute for SP data. The offset panel (if installed) provides position and elevation data to the NC during the mission. This data is used to define the location of selected radar fixpoints (offset aimpoints) relative to a target or destination.

A Ballistics Computer Unit (BCU), located in the right equipment bay, contains the electronics and electro- mechanical computing components to automatically compute weapon trail and time of fall. Lead Computing Optical Sight pipper azimuth and elevation positioning signals are also furnished to continuously sight on the weapon impact point for the existing conditions of airspeed, altitude, attitude, wind and ground-speed. Ballistics computations are accomplished only when the Nav computer is in a bomb mode.

MK II AVIONICS SYSTEM F-111D/F-111F/FB-111A DIGITAL BOMBING-NAVIGATION SYSTEM

The revolutionary Mark II system, ordered in June 1966, was made up of more than 75 main components. These components included: Inertial Navigation Set and Attack Radar, produced by North American Rockwell's Autonetics Division (General Dynamics' subcontractor for the complete Mark II system); Computer, International Business Machines' Federal Systems Division; Converter and Panels, Kearfott Division of Singer-General Precision, Inc.; Integrated Display Set, Norden Division of United Aircraft Corporation; Doppler Radar, Commercial Products Division of Canadian Marconi Company; Horizontal Situation Display, Astronautics Corporation of America; and Stores Management Set, Fairchild Hiller Corporation's Space and Electronics Division.

The F-111s Digital Bombing and Navigation system is made up of an aided-inertial digital system consisting of the Inertial Navigation Set (INS), the Control and Display Set (CDS) and the Digital Computer Complex (DCC) The system operated in conjunction with other weapon system avionics.

The Inertial Navigation Set (INS) is a fully automatic, self-contained inertial navigator which contains the Navigation Computer Unit (NCU), Inertial Reference Unit (IRU), and the Battery Unit (BU). The NCU is a self-contained digital navigation computer providing basic navigation computations when supplied inputs of velocity and heading. The NCU also provides the control and self-test functions for the inertial reference unit. The IRU contains a four gimbal, all attitude inertial platform and its associated electronics. The IRU provides basic outputs of velocity heading, and aircraft attitude. The BU provides electrical power control for the INS. The battery unit provides power for periods up to five seconds, for the NCU and IRU during aircraft electrical system transients.

The Control and Display Set (CDS) consists of three units, the Navigation Data Display Panel (NDDP), the Flight Data Panel, and the Navigation Data Entry Panel (NDEP). The control and display set controls primary electrical power to the avionics of the Bomb Nav system, and Doppler radar. It is used for entering, changing, and recalling digital computer data, selecting functional and operational modes, and display navigation, weapon delivery, and operational status data.

The Digital Computer Complex (DCC) is located in the forward electronics bay and is made up of the General Navigation Computer (GNC), the Weapon Delivery Computer (WDC), and the Converter Set (CS). The GNC is programmed to handle all navigation functions, and has a weapon delivery backup capability. The WDC performs weapon delivery control, provides system self-test, and serves as backup for navigation modes. The CS acts as a buffering logic and signal conversion device between the computers and other aircraft subsystems.

ATTACK RADAR

F-111A/F-111C/F-111E (AN/APQ-113)

The attack radar provides all weather navigation air-to-ground and air-to-air attack capability. Basic components of the radar set consist of an antenna assembly, an antenna roll unit, and an antenna control, all located in the nose radome; a modulator-receiver-transmitter (MRT) and an electrical synchronizer, located in the forward electronics equipment bay; and a radar scope, radar control panel, and a tracking control handle, all located at the WSO station.

F-111D (AN/APQ-130)

The attack radar system (ARS) provides all weather operation for aid to navigation, weapon delivery and reconnaissance. The system detects and tracks air and ground targets, provides long range ground mapping, operates with air-to-ground beacons and directional air-to-air IFF, provides for Continuous Wave missile illumination, and reports failures in-flight through built-in tests and continuously performed self tests.

The multi-sensor display (MSD) panel located on the WSO's instrument panel, is a primary control and display for the attack radar. The MSD screen is similar to a standard television cathode ray tube. The panel is interfaced with other avionics systems and displays TFR and the attitude director indication information similar to that being displayed on the vertical situation display (VSD). A simplified attitude director indicator presentation is available on the MSD simultaneously with TFR and attack radar presentations if desired. In addition, an antenna tilt scale, radar cursor and/or designation cursor, and readouts of latitude, longitude, heading, and radar range are displayed with each applicable mode.

F-111F (AN/APQ-161 and AN/APQ-169)

The attack radar provides all weather radar sensor operation to aid the navigation, bombing, and air-to-air search/track capability The set performs ground mapping, radar bombing, radar position fixing and correction, altitude calibration, air-to-ground ranging (with TFR), Airborne Instrument Low Approach (AILA)air-to-air search and tracking, weather avoidance, tanker location, ship surveillance, air/ground beacon, radar cueing for Pave Tack IDS and weapons, infrared sensor controls and video display, date/time video display, and weapons video display.

FB-111A (AN/APQ-114)

The attack radar provides all weather operation for aid to navigation, bombing an air refueling rendezvous capability. The set performs ground mapping, navigation fix-taking, air-to-air search and tracking.

F-111B

The F-111B Phoenix Firepower Control System is discussed in the F-111B chapter.

TERRAIN FOLLOWING RADAR

The terrain following radar (TFR) provides low altitude terrain following, obstacle avoidance and blind letdown capability. The TFR consists of left and right antenna receivers, synchronizer transmitters. power supplies and computers in a dual channel configuration, and a control panel. The vertical situation display (VSD) functions in conjunction with the TFR to provide PPI and E scope displays of the different modes of TFR operation. Each channel may be operated independently of the other in any one of three modes; terrain following (TF), situation display (SIT), or ground mapping (GM). The TFR receives input from the radar altimeter, attack radar, converter set, flight controls system, Doppler radar, central air data computer and Auxiliary Flight Reference System. The TFR operates on 115 volt ac power from the main ac bus and 28 volt dc power from the main dc bus.

The following are the designations of the F-111 TFR types by aircraft type:

F-111A/F-111C/F-111E	(AN/APQ-110)
F-111D	(AN/APQ-128)
F-111F	(AN/APQ-146)
FB-111A	(AN/APQ-134)

This WSO's eye view of a practice bombing target on the Nellis AFB range shows the outstanding forward visibility from the F-111 cockpit. (Author)

After the upgrade programs, the aircraft used the AN/APQ-171.

ARMAMENT SYSTEM

The armament capability of the aircraft F-111 includes the delivery of nuclear and conventional weapons, including guided weapons, in various configurations and either air-to-ground gunnery or air-to-air missiles. Manual air-to-air gunnery is also available; however, this mode is not computerized. Bomb carriage and launching equipment (pylons and stores release system), weapons bay doors, the weapons themselves, and (F-111A, F-111D, F-111E, and F-111C and F-111F when the Pave Tack pod is not installed) the M61A1 gun are considered as the armament system. Initially all Tactical F-111s could carry many different stores (33 conventional weapons, 3 nuclear bombs, fuel tanks, and two types of electronic countermeasures pods the QRC-160-8 and the QRC-335-4). In addition RAAF F-111Cs have been modified to carry AGM-84 Harpoon and AGM-142 Have Nap missiles.

BOMBING AND LAUNCHING EQUIPMENT

Bombing and launching equipment consists of the various bomb racks, stationary and pivoting wing pylons, and the release systems. The stores can be carried in the weapons bay and on eight wing pylons. Four of the wing pylons pivot to remain streamlined with different positions of the wing. The pivoting pylons are utilized for stores carriage in various configurations.

AVIONICS MODERNIZATION UPGRADE PROGRAM (F-111A/E) and PACER STRIKE (F-111F)

The older analog avionics of the F-111A/E, FB-111A and F-111Fs were upgraded as part of the Avionics Modernization Program (AMP) for the F-111E, and Pacer Strike for the F-111F. The upgraded avionics included a new ring laser gyro INS, Global Positioning System (GPS), and a new digital computer. Two multifunction displays (MFDs) added to the right

F-111F 72-1450 is seen carrying two GBU-15s, an ALQ-131 pod on the forward station and an AXQ-14 Data Link pod on the aft station. (Jim Rotramel)

CHAPTER TWO: THE F-111 DESIGN

side instrument, along with a Control Display Unit (CDU) on the right console were used by the WSO to input and modify data.

The AMP FB-111A and resulting F-111Gs had the attack radar and TFR replaced with the newer APQ-169 and APQ-171. The Australian Avionics Update Program (AUP) used the USAF AMP and Pacer Strike as a basis to develop an advanced update, installing the APQ-169 and APQ-171 in the F/RF-111Cs.

PAVE TACK/DATA LINK SYSTEM (F-111F and F-111C)

The Pave Tack/Data Link Systems consists of two major systems; the AN/AVQ-26 Pave Tack Guided Weapons system and the AN/AQX-14 Data Link System. The purpose of the combined system is to provide an improved target acquisition capability for launching and guiding weapons equipped with laser, television, or imaging infrared (IIR) seeker heads. The system provides stand-off delivery capability improves navigation accuracy. The Pave Tack video also provides bomb damage assessment data.

PAVE TACK SYSTEM

The Pave Tack system consists of the Electro-Optical Target Designator Set, AN/AVQ-26 (Pave Tack pod). control panels, visual monitor, and other components required to provide the circuitry necessary to launch and guide laser, TV, and IIR guided weapons. It provides weapon delivery and navigation capability through infrared sensing and laser ranging and designating. The infrared sensor makes possible the detection and tracking of targets during day or night. The laser rangefinder/designator provides for designation of targets for precision delivery of laser guided weapons and provides accurate slant range information for conventional weapon delivery and for accurate Bomb/Nav system updates.

The pod attaches to the Pave Tack cradle which is installed in the weapon bay. The cradle rotates 180 degrees to extend the pod for operational use or to retract it into the

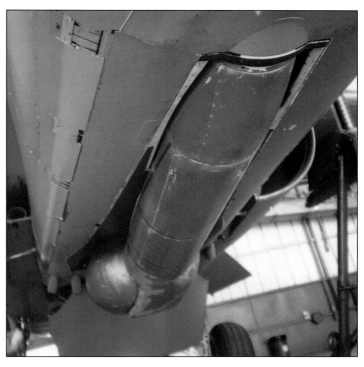

(Jim Rotrotramel)

(Jim Rotrotramel)

weapon bay. Rotation is accomplished by a hydraulic motor and mechanical drive mechanism. Operation of the Pave Tack cradle is inhibited through the main landing gear squat switch when the weight of the aircraft is on the wheels and the weapon bay doors are open. With the doors open, the cradle ready switch located in the main wheel well must be actuated to operate the cradle during ground operations. The pod consists of two basic sections, the head and the fixed base. The pod head section integrates the infrared and laser sensors with a movable optic system providing complete IR or laser line-of-sight coverage of the hemisphere below the aircraft. The pod head is stowed to protect the optical window when the pod is retracted into the weapon bay, when the pod is

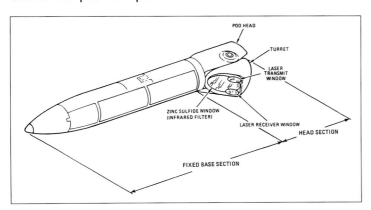

29

rotating, or when pod power is in standby. The head extends by rotating 90 degrees when the standby mode is exited and a search or track mode is entered. The laser can be fired only when the head is extended. The pod fixed base section contains the power supplies and avionics to operate and control the Infrared Detecting Set (IDS) and the laser. The mechanism required to drive the pod head is located in the base section. The Tracking Control Unit (TCU) is used to control the pod line of sight and position the Pave Tack cursor for fix-taking, bombing, and target tracking. The Virtual Image Display (VID) displays a TV video of the pod field of view. The VID also displays pod status alphanumeric data. Self-test features incorporated in the Pave Tack system are used for preflight, inflight, and maintenance malfunction analysis and troubleshooting. The system has five search modes and two track modes.

In order to guide the GBU-15 an AXQ-14 data link pod seen here must be used. The pod can also guide weapons released from other aircraft. (C. Roger Cripliver Collection)

DATA LINK SYSTEM

The Data Link system consists of the AN/AQX-14 Data Link pod (Television Receiver-Transmitter), control panels, visual monitor and other components required to provide the circuitry necessary to launch and guide TV guided weapons. It provides the capability for accurate delivery of television guided weapons. In-flight remote control of the weapon is provided by the Data Link pod. The primary components of the Data Link system are the Data Link pod, the guided weapons, and at the WSO station, the Data Link control panel. The system uses the same Tracking Control Unit (TCU) and Virtual Image Display Group (VID) as the Pave Tack System. The Data Link pod is attached to the aircraft aft centerline station with a pod pylon and is directly operated by the WSO. The pod receives seeker video from the guided weapon and transits the video for display on the VID. Simultaneously, commands from the WSO are transmitted by the pod to the weapon. The weapon video is received, processed, and supplied to the VID. The Data Link guided weapons can be gen-

The F-111A in the Pave Mover program, 66-0053, shows off the Norden/Grumman radar fairing is visible in this photo. (USAF via Brian C. Rogers)

CHAPTER TWO: THE F-111 DESIGN

F-111E 67-0115 with the Hughes Electronics version of the Pave Mover radar had a black fairing as seen here. (USAF via Marty Isham)

F-111D 68-0085, seen here with Eglin AFB in the background, used a white fairing to cover its Hughes Electronics radar. (USAF via Brian C. Rogers)

erally categorized as Guided Bomb Units (GBU) and air-to-ground missiles(AGM). The weapons control panel is used to select the guided weapon type, station, fuzing, and delivery mode. The weapon is locked-on to the target and steered by the WSO. Video from the weapon is displayed on the VID.

PAVE MOVER PROGRAM

During 1982 and 1983, three F-111s; one F-111A, one F-111D and one F-111E were used in the testing of a battle field surveillance Synthetic Aperture Radar (SAR) system being developed under the Pave Mover program. The Pave Mover program used a Target Acquisition/Weapon Delivery System (TAWDS). The purpose of the radar was to build a radar map of the battlefield area in order to locate enemy ground targets in real time and then direct friendly aircraft to attack these ground targets. Two versions of the radar were tested. The Norden/Grumman design was housed in a canoe under the belly which resembled the canoe of the EF-111A. The other design was developed by Hughes Electronics.

CHAPTER THREE

U.S. Air Force Tactical F-111s

F-111A

The first Pre-production F-111A, 63-9766 was rolled out of the General Dynamics' Fort Worth plant on October 15, 1964, 37 months after Secretary McNamara's go-ahead decision, 22 months after the program's actual beginning, and two weeks ahead of schedule. The October 15 roll-out ceremonies prompted McNamara to remark: "The Air Force, the Navy, and General Dynamics and its subcontractors ... have produced a plane which will fly faster at any altitude than our best current fighter – a plane with several times the payload and twice the range of any previous fighter-bomber. One F-111 will have the fire power of five World War II Flying Fortresses. For the first time in aviation history, we have an airplane with the range of a transport, the carrying capacity and endurance of a bomber, and the agility of a fighter pursuit plane."

The first flight was made from Carswell AFB, Texas, on December 21, 1964. The overall results were satisfactory, although the flight was shortened to 22 minutes because of flap malfunctions, and engine compressor stalls. The aircraft immediately entered Category I testing. During this early testing period, the F-111A achieved Mach 1.3. In addition as planned, maintenance proved to be comparatively simple. On February 25, 1965, the maiden flight of the second F-111A, the F-111 swept its wings from a 16" to a 72.5" aft position (as designed). These first two aircraft were the only test F-111s accepted by the Air Force (each on its first flight's date) prior to the initial production agreement.

Flight tests of the F-111A continued from 1964 through 1973. In 1965, the flight test program was expanded, adding eleven F-111A production aircraft (65-5701 through 65-5710, and 66-0011) to the first 18 RDT&E F-111As. The Category I flight tests which started in December 1964 did not end until March 31, 1972. At that time, Category II tests which had begun in January 1966, were still going on. Several postponements slipped the Category III tests to 1969. They were finally canceled as operationally unnecessary.

The F-111 had some initial problems. Engine malfunctions and weight increases were the main problems. This was not unusual during the development of high-performance aircraft, even those less revolutionary than the F-111A. The Pratt and Whitney P-1 engine, the production version of the afterburning turbofan TF30, was first flown in an F-111A on July 20, 1965. Despite thorough testing of the experimental TF30, problems were soon detected. The first 30 F-111As (each equipped with two P-1 engines) had numerous engine stalls, particularly at high Mach speeds and high angles of attack. The rest of F-111As received the P-3, an improved P-1 which had become available in 1967. The new engine was later retrofit in several of the first 30 F-111As. It was installed with an air diverter (Triple Plow I intake). The P-3/Triple Plow I combination did not totally cure the stall problem, but it did help. The modification required few airframe changes and led to further progress on engine intake improvements (the Triple Plow II Improved Intake System was installed on other USAF F-111 models). The production F-111A's takeoff weight for conventional missions (92,000 lb) exceeded the DoD September 1961 specification by 30,000 lb, but exceeded USAF expectations by 10,000 lbs.

The first production F-111A (65-5701) first flew on February 12, 1967. By August the Air Force had accepted these two and nine others, committing the eleven aircraft to the test program. Even with the engine problems, an RDT&E F-111A reached the design top speed of Mach 2.5 on July 9, 1966.

On May 1, 1967, an F-111A set a flight record of 7 hours and 15 minutes without refueling. On May 22 of the same year, two F-111As attained a fighter-type aircraft unofficial record for transatlantic flight without refueling or external tanks. The two (on their way to the Paris Air Show) flew from Loring AFB, Maine, to Le Bourget Airport, Paris in 5 hours and 54 minutes. They covered 2,800 nautical miles at an average speed of 540 mph, flying with their wings extended in cruise position most of the time.

Combat Bullseye tests were conducted at Nellis AFB during April and May 1967. The purpose of these tests was

CHAPTER THREE: U.S. AIR FORCE TACTICAL F-111s

66-0044 is seen here in the "pitch out" for landing at Nellis AFB. (Author)

to prove the weapon delivery capabilities of the F-111A. The Combat Bullseye I tests had immediate impact. They confirmed the superior bombing accuracy of the aircraft's radar and convinced the USAF that the F-111A could be deployed to Southeast Asia.

The first flight of an F-111A (66-0013, the 31st production), featuring the P-3 engine/Triple Plow I air diverter combination was successfully conducted on September 24, 1967.

Six of the first operational F-111As (66-0017, 66-0018, 66-0020, 66-0021, 66-0022, and 66-0023) began arriving at Nellis AFB, Nevada on July 17, 1967, to be used as part of the initial air crew training program known as Combat Trident. Combat Trident flew 2,000 flying hours and 500 bombing sorties to train the crews, which completed on March 6th, only nine days before the Combat Lancer deployment to Takhli Royal Thai Air Force Base (RTAFB). The F-111 aircrew was initially made up of two pilots – the aircraft commander (AC) in the left seat, and the Pilot-Weapon Systems Officer (P-WSO) in the right seat. Starting around 1970, following the lead of the USAF F-4s, P-WSOs were replaced with Navigators specially trained as a Weapon Systems Officers (WSOs). A few P-WSOs continued in the crew force, but the majority of right seats in the F-111s were filled by Navigators.

The 474th TFW at Cannon AFB, New Mexico, had given up its F-100s and moved to Nellis AFB, without personnel or equipment on January 20, 1968. The wing initially had two squadrons, the 4527th Combat Crew Training Squadron (CCTS) which was activated on January 20, and the 428th Tactical Fighter Squadron (TFS) which was redesignated from the 4481st on the same day. As more aircraft were delivered, two additional squadrons were activated; the 429th TFS activated on May 15, 1968 and the 430th TFS activated on September 15, 1968.

The 428th TFS reached an initial operational capability on April 28, 1968 at Nellis AFB. Due to the Harvest Reaper modifications validated by Combat Lancer operations, and other unexpected modifications required to be added to the F-111As, the wing was not operationally ready until July 1971.

The early modified F-111As at Nellis were known as the Harvest Reaper jets (66-0013 through 66-0025), and they could be distinguished from the other early F-111As by their FS 34079 dark green underside. The normal production aircraft at the time had the standard SEA camouflage FS 36622 light grey on the underside. The Harvest Reaper modifications consisted mainly of improvements/additions to the avionics and electronic countermeasures (ECM) equipment. On March 15, 1968, only eight months after the first aircraft had arrived at Nellis AFB, a group of the Harvest Reaper F-111As deployed to Takhli RTAFB, Thailand, as Operation Combat Lancer. Under Combat Lancer, a small detachment of six F-111As (66-0016, 66-0017, 66-0018, 66-0019, 66-0021, and 66-0022) deployed to take part in the war in Southeast Asia. After the loss of two aircraft (66-0017 and 66-0022), two replacement aircraft (66-0024 and 66-0025) deployed to Takhli, arriving on April 1, 1968. 66-0024 was lost on April 22, and combat operations were suspended the same day. The remaining six aircraft returned to Nellis AFB on November 22, 1968, after flying 55 combat missions during the Combat Lancer deployment. (See Combat Lancer in the COMBAT OPERATIONS section).

During 1969, improvements to problems discovered during flight operations and Combat Lancer were incorporated as part of Harvest Reaper, and planned to be added to new aircraft during production. The Harvest Reaper improvements, although approved for production in April 1968, were delayed. The Air Force decided that the improvement program should include modifications called for as a result of a three month evaluation of Combat Lancer. The evaluation ended in August but it took longer than expected to tie Combat Lancer crashes to a malfunction of the aircraft's tail servo actuator (caused by a tube of sealant lodged in the actuator) in one case, poor mounting of the M-61 gun, and pilot error in the two others. Similarly, F-111 testing and training incidents, including two crashes in early 1968 (63-9769 and 65-5701) dictated a detailed evaluation that became quite involved. On August 27, 1969, (1 day after the beginning of the F-111A's Category II fatigue tests) an F-111 wing carry-through-box failed during a ground fatigue test.

The beginning of the F-111A's fatigue test program slipped from February 1965 to July 1968. Design and weight reduction changes had to be incorporated into the test airframe to assure realistic testing; and also, General Dynamics' late submission of acceptable testing procedures caused the testing to be delayed. A final three month delay was caused by late modifications to incorporate the new Triple Plow I air diverter, correct a deficient wing carry-through-box (that had failed during early static tests), and fix an unsatisfactory tail pivot shaft fitting.

In early 1969 General Dynamics discovered that Selb Manufacturing, who made the defective steel wing carry-through-boxes, was paying off inspectors for approving unauthorized weldings. An FBI investigation followed. A Federal Grand Jury indicted General Dynamics in 1972, for destroying $114,000 worth of flawed boxes and filing a claim with the Air Force for repayment, instead of charging the loss to Selb. A trial jury acquitted General Dynamics in 1973.

As a result, General Dynamics' overall improvement of the F-111 (particularly, additional Harvest Reaper avionics) did not go ahead as planned. The updates started in January 1969, and required extensive retrofit because most F-111As had already been delivered. If possible when necessary, the retrofit modifications were integrated into the production of later F-111s.

Early F-111A and F-111E aircraft had deficient windshields. On May 29, 1969, an F-111A (67-0043) on a training flight at Nellis crashed at low altitude when the windshield bulged in down from the top of the canopy bow and instantly crazed. Unprotected from the wind, with shards of windshield glass flying around the cockpit, the crew could not see, communicate, or control the aircraft and were forced to eject. TAC replaced 50 F-111 windshields in 1969; and 93, the following year. However, this did not solve the bird-strike problem, shared by all F-111s and other older high-speed aircraft. By mid 1973, 52 F-111s suffered damage from bird strikes resulting in two more F-111s (66-0029 and 67-0040) being lost. All three of these aircraft remained airworthy prior to crashing. Such losses, in 125,000 flying hours, dictated that something had to be done about the problem.

This reaffirmed the urgent need for a stronger windshield. TAC wanted one that could withstand the impact of a four pound bird at 500 knot airspeed, but the exorbitantly high costs killed this proposal. In mid-1973, development of an improved, reasonably priced windshield still showed little progress. It was the following year before a contract was let, and testing consumed another year. Meanwhile, the Air Force tested a Navy helmet that promised some windblast protection because of its polycarbonate faceplate. The Air Force helmets at that time used an acrylic faceplate. Individual helmet liners (foamed-fitted to the pilot's head) were obtained. They helped considerably in preventing crews from losing their helmets if their windshield broke. The Air Force also continued evaluating strobe lights to reduce bird strikes. Fifty F-111s took part in the program.

The Air Force lost 67-0049, its 15th F-111 loss, on December 22 1969, due to failure of the forged wing pivot fitting (a part of the basic wing structure, located next to the wing-carry-through-box). This accident triggered renewed criticism of the aircraft. In Congressional testimony on March 17, 1970, the Secretary of the Air Force admitted difficulties but pointed out, "This plane per thousand hours flown, has fewer accidents than any other Century Series aircraft." In February 1972, after 150,000 hours, the F-111 still had the lowest accident rate of the nine most recent USAF/USN high-performance tactical aircraft, even though a large percentage of its flights were low level (200' to 500' above the terrain), and mostly at night. The F-111 accident rate in early 1972, was 40% under that of the F-106, USAF's next safest aircraft. However, the loss of the 15th aircraft caused the grounding of all F-111s the next day (December 23, 1969). The few F-111As used in flight testing continued to fly.

The December 1969, accident cast doubt on the F-111's structural integrity. The January 1969, improvement program (including the delayed addition of Harvest Reaper avionics) had already been expanded to include wing-carry-through-box structural modifications which extended the fatigue life to the 10-year contractual design requirement. Cyclic loads ground testing of a modified wing-carry-through-box were resumed in December 1969. They gave the box a test-life of 24,000 hours (equivalent to a safe service-life of 6,000 hours).

With Air Force authorization, General Dynamics, on May 18, 1970, gave North American Aircraft a development contract for a new titanium wing-carry-through-box .

Investigation of the December 22, 1969, F-111A (67-0049) crash dictated a thorough structural inspection and proof testing program which was called the Recovery Program. This Program was a $31.2 million, non-destructive, cold-proof testing and modification effort, started in the spring of 1970. The grounding was lifted on July 31, 1970.

The Recovery Program's $31.2 million amount would cover nonrecurring costs for materials and equipment, plus the recurring costs for labor to take each aircraft through inspection and testing. The Air Force wanted General Dynamics to accomplish this under the contract's correction of deficiencies clause. Approved aircraft procurement took care of inspection and proof testing funding. The Air Force covered the expense by dropping several F-111Fs from a follow-on buy.

The Air Force believed that blending this project with the F-111's overall modernization, would restore the F-111s to operational status in early 1971. Very little slippage occurred. TAC input the first F-111A scheduled for update to General Dynamics in April 1970, and by December 1971, the last of 340 F-111s (including 125 F-111As) had been processed. The Recovery testing of each F-111 covered more than a dozen structural components, four of which required load-proof testing at a temperature of minus 40 degrees F. A few bolts broke, which was not surprising, yet no forging defects appeared. Still cautious, in August 1971, the Air Force scheduled further (Phase II) structural in-house inspections of every F-111 model. Each F-111A had to undergo Phase II inspections before reaching 1,500 flying hours. F-111E and D aircraft fell under the 1,500 flying hour criterion. The F-111F and FB-111A could log up to 2,000 and 2,500 flight hours, respectively, prior to undergoing Phase II inspections. The first F-111A entered Phase II processing (II Structural Integrity Program(II SIP)) at the Sacramento Air Materiel Area (McClellan AFB) on May 16, 1973.

In early March 1972, F-111As from the 474th TFW at Nellis AFB prepared for their second deployment to Southeast Asia. On March 15, as part of Constant Guard V, the 429th TFS and 430 TFS deployed to Tahkli RTAFB, the base used during Combat Lancer. They continued to support the air war with night time raids deep into North Vietnam. By the end of combat operations in mid 1973, the F-111As had flown over 4,000 combat sorties, losing only six aircraft for a loss rate of 0.015%, the lowest rate of any combat aircraft in the Vietnam War. (See Constant Guard/Linebacker in the COMBAT OPERATIONS section).

This is a pilot's eye view of flying as "blue" four in a four ship formation (Author)

After return from Southeast Asia in the summer of 1975, the F-111As continued to be assigned to the 474th TFW and operate from Nellis AFB. The 474th TFW was once again at four squadron strength, being made up of the 428th TFS, 429th TFS, 430th TFS, and 442nd Tactical Fighter Training Squadron (TFTS), (which had replaced the 4527th CCTS on October 15, 1969).

During 1976 three F-111As were lost due to engine failures. The problem was traced to a bad seal at the base of the engine turbine blades, (a similar problem had shown up on the F-111F's P-100 engine). After a grounding of almost three months, a fix was found and the aircraft were returned to flying status.

The F-111As transferred to Mountain Home AFB, Idaho, joining the 366th TFW, as part of Operation Creek Swing/Ready Switch, in August 1977. This occurred because the United States Air Forces Europe (USAFE) needed the night/low level strike capability of the F-111F. The operation transferred F-4Ds from Lakenheath Air Base in the United Kingdom to replace the 474ths F-111As at Nellis AFB. The Nellis AFB F-111As went to Mountain Home AFB, which allowed the F-111Fs to move from Mountain Home to Lakenheath. Starting in the late 1970s, many of the F-111As flying at Mountain Home AFB were assigned to a training role.

In 1975, the first of 42 of the F-111As were modified to EF-111A configuration, with the last being completed in December 1985. All remaining F-111As were retired by the spring of 1992.

The total of 159 F-111As accepted included 18 RDT&E F-111As ordered in December 1962, even though the 18th test F-111A (63-9783) was used as an FB-111A bomber prototype and charged to the SAC program. Four RDT&E F-111As were accepted in fiscal year (FY) 1965, eight in FY 1966, and five in FY 1967. The Air Force accepted five F-111A production aircraft in FY 1967, 36 in FY 1968, 86 in FY 1969, and 14 in FY 1970. Monthly acceptances averaged three F-111A aircraft until July 1968, when they rose to seven.

The total cost of Research, Development, Test and Evaluation (RDT&E) was $1.657 billion, $200 thousand more than estimated by General Dynamics, and still $1.176 billion over the target cost of May 1964. The cost of each F-111A was $8.2 million. This included; airframe – $4,304,000; engines (installed) – $1,354,000; electronics – $1,688,000; ordnance – $7,000; and armament – $925,000. This cost was excluding some $2.8 million spent for RDT&E and about $800,000 worth of modification, bringing the actual cost of each F-111A to more than $11.8 million. By mid 1973, the average cost per flying hour was $1,857.00.

PRE-PRODUCTION and TEST F-111As

The first 30 F-111As, factory numbered A1-010 through A1-30, were Research, Development, Test & Evaluation (RDT&E) aircraft. The 1963, 1965, and the first two 1966 year F-111A aircraft were used in RDT&E testing. Contrary to reports in other publications, according to General Dynamics, there were no YF-111As built; all 30 of the RDT&E aircraft were F-111As.

The first 30 flew in Category I (Contractor) Tests, Category II (USAF) Tests, and Category III (Operational) Tests. The Category I tests were carried out at Carswell AFB (site of General Dynamics Fort Worth factory) and at Edwards AFB, Category II tests primarily at Edwards AFB, and Category III tests primarily at Nellis AFB and Eglin AFB. Many of the surviving RDT&E airframes were later used as spare parts for repair of operational F-111s.

63-9766 (A1-01)
The first F-111 was accepted by the USAF and made its maiden flight on December 21, 1964. It was piloted by General Dynamics test pilots Dick Johnson in the left seat with Val Prahl in the right seat. It was delivered in a gray over white paint scheme and was fitted with ejection seats. The first supersonic flight occurred on its ninth flight, reaching Mach 1.2 on March 5, 1965. It hit Mach 2.03 on August 8, 1965. It last flew on August 5, 1967, and was retired from flight duties after 210 missions with a total of 253 flight hours, 18 hours and 43 minutes of which were at supersonic speeds. It was used from September 1, 1967, to February 28, 1968, for refused take-off and barrier engagement testing. The aircraft is preserved in the outdoor airpark, part of the Air Force Flight Test Center Museum, at Edwards AFB, California. (Lockheed Martin Tactical Aircraft Systems)

CHAPTER THREE: U.S. AIR FORCE TACTICAL F-111s

63-9767 (A1-02)
Originally a bare metal aircraft with ejection seats, 63-9767 was a test bed for Mk.1 NAV/ATTACK system. Its first flight occurred on February 25, 1965, and was accepted by the USAF on the following day. In 1966 it was delivered to Pratt & Whitney to be used as an engine/intake testbed. In October 1965, the YTF30 engines were replaced with TF30 engines. On July 1, 1967 the aircraft was transferred to Bradley Field, East Hartford, Connecticut. It continued to fly engine test flights for Pratt & Whitney until returning to GD, Fort Worth on March 4, 1968. Its last flight, to Chanute AFB, occurred on December 4, 1969. 63-9767 had accumulated a total of 467.8 flight hours during 320 test flights. It retired to Chanute AFB, and is preserved at the Octave Chanute Aerospace Museum, Rantoul, Illinois. (Lockheed Martin Tactical Aircraft Systems)

63-9768 (A1-03)
This aircraft was a pre-production test aircraft with ejection seats. It first flew and was accepted by the USAF on April 30, 1965. Its last flight, to Sheppard AFB, occurred on December 4, 1968. It was used for ground training at Sheppard AFB, Texas. 63-9768 had accumulated 564.2 flight hours during 300 test flights. During 1995 the aircraft was transported by truck to Norfolk VA, then by sea aboard HMAS Tobruk to Sydney and then to RAAF Amberley to be used for Battle Damage Repair (BDR) ground training. It is currently undercover at Amberley, but in poor condition. (Lockheed Martin Tactical Aircraft Systems)

63-9769 (A1-04)
The aircraft was originally bare metal and had ejection seats. It first flew and was delivered to the USAF on June 29, 1965. A spin-recovery parachute was added for use during Research Development Testing and Evaluation (RDT&E). It was the first F-111 with working pivot pylons. It performed a large number of the early aerodynamic and stores compatibility trials, and flew with mock up AIM-54 Phoenix missiles. The aircraft stalled, crashed and was badly damaged on its 229th flight during a low speed flyby during an airshow at Holloman AFB. The crash occurred May 18, 1968. The crew did not attempt ejection and was not injured. When the aircraft crashed, it had accumulated 420.2 flying hours. It was scrapped on August 7, 1968 at Holloman with the usable components being sent to Fort Worth as spares. (Lockheed Martin Tactical Aircraft Systems)

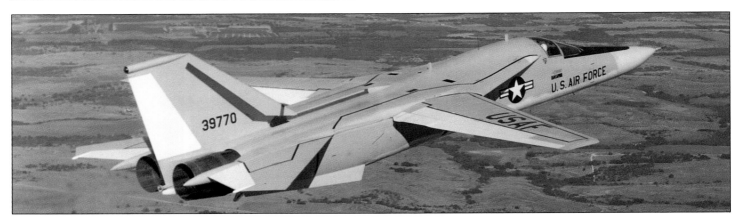

63-9770 (A1-05)
This aircraft first flew and was accepted by the USAF on July 31, 1965. It was used to test the M61A1 Vulcan 20mm Gattling gun, F-111D AIM-7D installation, and electronic warfare systems, flying many of its missions from Eglin AFB, Florida. It was equipped with ejection seats. Its last flight, to Sheppard AFB, Texas occurred on April 1, 1969. It was used for ground training at Sheppard AFB. It had accumulated 396.6 flying hours during 222 flights. (Lockheed Martin Tactical Aircraft Systems)

63-9771 (A1-06)
This pre-production first flew and was delivered to the USAF on December 24, 1965. 63-9771 had ejection seats. It was used by NASA at the Dryden Flight Research Facility (DFRC), Edwards AFB, California. After 148 flights, it was retired to Military Aircraft Storage and Disposition Center (MASDC), (now called AMARC) on December 18, 1969, with 284.9 flying hours. On June 6, 1972, it was shipped by C-5A to Cannon AFB, New Mexico where it is still preserved. The aircraft was first marked as tail number 27-234, then just 27 (27th FW), and now 63-9771. (Lockheed Martin Tactical Aircraft Systems)

63-9772 (A1-07)
This pre-production test aircraft had ejection seats. It first flew and was delivered to the USAF on October 19, 1965. It was used in weapon separation and missile launch testing, including GAR-8/AIM-9 launches using the weapons bay mounted trapeze. It was slightly damaged on a weapons effectiveness test flight on January 26, 1968. After 199 flights, it was retired on March 21, 1969, with a total of 226.4 flight hours, and later used for ground training at Sheppard AFB. (Lockheed Martin Tactical Aircraft Systems)

CHAPTER THREE: U.S. AIR FORCE TACTICAL F-111s

Right: This photo shows the weapons bay missile trapeze used for the launch of GAR-8/AIM-9 air to air missiles. (Author)

63-9773 (A1-08)
Below: This pre-production test aircraft was used in Inflight Refueling Test. It first flew on December 30, 1965, was accepted by the USAF on February 16, 1966. The aircraft had ejection seats, and after 225 flights, was retired to Sheppard AFB on November 19, 1969 with a total of 508.0 flying hours. At Sheppard it was used for ground training. It is now on display at Sheppard AFB. (Lockheed Martin Tactical Aircraft Systems)

63-9774 (A1-09)
This pre-production test aircraft first flew on January 29, 1966, and was delivered to the USAF on February 16, 1966. It was fitted with ejection seats. The aircraft was the first F-111 lost, crashing on January 19, 1967, on its 108th flight, at Edwards AFB. The landing was planned for 16 degree wingsweep; the wings were actually at 50 degrees, with the inlet cowls closed. This caused the aircraft to hit the ground 1 1/2 miles short of the runway. The pilot, Major Herbert F. Brightwell, though not significantly injured by the crash, was killed when engulfed by fire while attempting to free the WSO, Donovan I. McCance, from his seat. The WSO survived, but was badly burned. The aircraft was badly damaged by the fire and was not repaired. It had a total of 194.3 flying hours when it crashed. (Lockheed Martin Tactical Aircraft Systems)

63-9775 (A1-10)
This pre-production test aircraft first flew on February 28, 1966, and was accepted on March 31, 1966, and had ejection seats. It deployed to Alaska during the fall of 1966, for cold weather testing. It was retired to Sheppard AFB on November 12, 1969, after 192 flights, with a total of 415.5 flying hours. It was used at Sheppard AFB for ground training. The aircraft is presently on display at U.S. Space and Rocket Center, Huntsville, Alabama. (Lockheed Martin Tactical Aircraft Systems)

63-9776 (A1-11)
This aircraft was built as an F-111A and delivered on March 28, 1966, It was modified to the RF-111A Prototype, the only RF-111A built. It first flew, as an RF-111A, on December 17, 1967. (See RF-111 Configurations) It was retired after 126 flights on April 24, 1970. It had accumulated 265.9 flight hours. The aircraft was the last F-111A with ejection seats. It is on display at Mountain Home AFB, Idaho, carrying tail number 66-0022 – the tail number of the first F-111 lost in combat. (Lockheed Martin Tactical Aircraft Systems)

63-9777 (A1-12)
Right: This was the first F-111 fitted with the ejection module, and first flew on May 27, 1966. It was delivered to the USAF on May 29, 1967. The aircraft was used in weapons testing by the USAF. NASA's Flight Research Center received the aircraft in April 1969 and was flown by NASA in a handling-qualities investigation program. After 178 test flights, accumulating 274.4 flight hours, the aircraft was retired to MASDC on July 29, 1971 and stored as FV008. (Lockheed Martin Tactical Aircraft Systems)

63-9778 (A1-013)
This pre-production test aircraft first flew on December 31, 1966, and was delivered to the USAF on May 29, 1967. It was used for G load testing. It was transferred on March 6, 1969 to NASA and used at the Dryden Flight Research Facility. Redesignated as a NF-111A, it was NASA's Transonic Aircraft Technology (TACT) Test Aircraft in 1973 (supercritical wing with reduced aspect ratio) and later, in 1986, was modified as the Advanced Fighter Technology Integration (AFTI) (with variable camber mission adaptive wing). The aircraft is now preserved at the NASA facility, Edwards AFB, California. (NASA)

63-9779 (A1-14)
This pre-production test aircraft first flew and was delivered to the USAF on July 20, 1966. During 1966, it was used to test the engine stall inhibitor system, which was designed to warn the pilot of an imminent engine stall. In 1967, It was used to flight test the Triple Plow II inlet. In 1970, it flew as a testbed for TF30-P-100 engines. After 199 test flights, it flew to General Dynamics, Fort Worth Texas, where it was used for parts. It had accumulated a total of 278.6 flight hours. The remaining hulk of the aircraft arrived at AMARC on February 12, 1990. (Lockheed Martin Tactical Aircraft Systems)

63-9780 (A1-15)
This pre-production test aircraft first flew and was delivered to the USAF August 21, 1966. It crashed and was destroyed October 19, 1967, at Boyd, Texas after speed brake support bracket failure resulted in a total hydraulics failure. With the total hydraulics failure, the aircraft became uncontrollable. The aircraft was on its 113 flight and had accumulated 278.6 flight hours. The General Dynamics crew was not injured in the ejection. The module is on display at the USAF Museum, Wright Patterson AFB, Ohio. (Lockheed Martin Tactical Aircraft Systems)

63-9781 (A1-16)
This pre-production test aircraft first flew on October 22, 1966, and was delivered to the USAF on November 20, 1966. It was transferred to Pratt & Whitney at Bradley Field, East Hartford, Connecticut on March 4, 1968, and was used for TF30-P-9 testing from April 20, 1968, through December 31, 1968. 63-9781 returned to GD, Fort Worth and flew as a chase and support aircraft until retired after 448 flights and 698.9 flight hours. It flew to MASDC on September 29, 1972. It was stored as FV012. (Lockheed Martin Tactical Aircraft Systems)

63-9782 (A1-17)
This pre-production test aircraft first flew on December 8, 1966, and was delivered to the Air Force on December 29, 1966. It was assigned to the USAF Flight Test Center (Air Force Systems Command), Edwards AFB, California on January 9, 1967. The aircraft was used for weapons testing. After 150 flights accumulating 272.4 flight hours, 9782 was put in storage at the MASDC, Davis-Monthan AFB, Arizona on March 3, 1970. On March 18, 1975, it was transferred to the Rome Air Development Center, (RADC), Griffiss AFB, N.Y. It arrived at Griffiss AFB during March 1975, aboard a C-5A Galaxy. On September 18, 1975, it was mounted on a pedestal for antenna measurement testing. The airframe, originally an F-111A, was reconfigured into an EF-111A shortly after its arrival at Rome Lab. In late 1975, it was reconfigured to an FB-111A through the addition of wing tip extensions, designed for use on short wing F-111s for ferry flights over long distances. These extensions made the F-111A wings the same length as the long FB-111A wings. Electronic counter measures pods and weapon data link pods have been evaluated using this F-111 test bed. As an EF-111A, the test bed was being used in the System Improvement Program (SIP). (Rome Air Development Center)

CHAPTER THREE: U.S. AIR FORCE TACTICAL F-111s

63-9783 (A1-18)
This modified RDT&E F-111A, while still equipped with TF30-P-1 engines and the tactical F-111A landing gear, served as FB-111A prototype. The aircraft flew for 45 minutes on its maiden flight on July 30, 1967, achieving Mach 2. Accepted by the USAF on August 18, 1967, it remained with General Dynamics for more testing. Category I testing, a prime contractor's responsibility, which started on July 19, 1967, and lasted through November, 1971. On August 13, 1968, the aircraft was damaged when hit by its own 600 gallon external fuel tank during jettison tests. The aircraft retired to MASDC (later renamed AMARC) on December 15, 1971, after 125 test flights accumulating 292.3 flight hours. (Lockheed Martin Tactical Aircraft Systems)

65-5701 (A1-19)
The aircraft first flew on February 12, 1967. It was delivered to the USAF on April 28, 1967. While on its 84th flight, over the Edwards AFB gunnery range on January 2, 1968, the aircraft experienced a weapons bay gun fire. The crew was forced to eject, with one crew member receiving back injuries. This was the first time the ejection module was used for an actual emergency. All other times were during unmanned development tests. The aircraft had a total of 207.7 flight hours when lost. (Lockheed Martin Tactical Aircraft Systems)

65-5702 (A1-20)
The aircraft first flew on April 27, 1967, and delivered to the USAF on the same day. It was used for stability and control testing, after which it was used for FB-111A/AGM-69A SRAM integration tasks. After flying 115 test missions and accumulating 237.8 flight hours. The disassembled aircraft was shipped to by truck to MASDC, Davis Monthan AFB, during September 1970. (Lockheed Martin Tactical Aircraft Systems)

65-5703 (A1-21)
This test aircraft first flew and was delivered to the USAF on August 24, 1967. It crashed and was destroyed on September 11, 1972, during Category II stall/post stall tests at Edwards AFB. A forty-foot diameter spin recovery parachute was contained in a canister attached to the tail of 65-5703, as seen in this photograph. On its 124th flight, during an intentional stall, the aircraft entered a spin at 35,000 feet. Deployment of the recovery parachute was initiated at 20,000 feet and 220-knots of airspeed. The parachute failed and separated from the plane. The crew ejected safely at 11,500 feet. Post crash investigation indicated the aircraft may have recovered to controllable flight, but the recovery was not recognized by the aircrew. The flight test spin recovery procedure had called for full aft stick to recover. After the loss, the procedure was changed to full forward and centered. The spin and the ejection was filmed by a camera in the cockpit. At the time I went through F-111 crew training, this filmed sequence was shown to all F-111 student aircrew members. The aircraft had accumulated 232.2 flight hours when it crashed. (Lockheed Martin Tactical Aircraft Systems)

65-5704 (A1-22)
This test aircraft first flew on October 25, 1967 and was delivered to the USAF November 9, 1967. It was configured and used to obtain Harvest Reaper performance data. In January 1970 it replaced 63-776 as a chase and support aircraft. It was retired to MASDC on March 11, 1971, after its 147th flight, having accumulated 321.8 flight hours, and stored as FV006. (Roy Lock)

CHAPTER THREE: U.S. AIR FORCE TACTICAL F-111s

65-5705 (A1-23)
Above: This test aircraft first flew on June 1, 1967 and delivered to the USAF on the same day. It was initially used for Category I flutter testing. In November 1967, it was modified for testing of F-111D Mk II Avionics Environmental Control System (ECS). 65-5705 retired to MASDC as FV005 on January 7, 1971. It had accumulated 99 flights totaling 127.3 flight hours. (Lockheed Martin Tactical Aircraft Systems)

65-5706 (A1-24)
Right: This test aircraft first flew on May 19, 1967, and was delivered to the USAF on July 20, 1967. 65-5706 was retired to MASDC on March 20, 1971, and then was shipped to General Dynamics, Fort Worth on March 20, 1972, to be used for spare parts. The aircraft had a total of 277 flights accumulating 603.2 flight hours. Some parts of the forward half of the aircraft were used to rebuild F-111E 68-0082. (Lockheed Martin Tactical Aircraft Systems)

65-5707 (A1-25)
The aircraft was delivered to the USAF on May 25, 1967 and assigned to the FB-111A test program. It was modified having FB-111A avionics installed and used in flight tests of the Mark IIB avionics system. Its first flight occurred on March 31, 1968. 65-5707 completed FB-111A avionics flight tests on September 4, 1970, and was placed in temporary storage at Edwards AFB. After 145 flights accumulating 472.3 flight hours. It was retired to MASDC on January 5, 1972, and was stored as FV009. (Mick Roth)

65-5708 (A1-26)
This test aircraft first flew on May 2, 1967 and was delivered to the USAF on May 31, 1967. It was used to test the full up ECM system, including the ALR-23 Countermeasures Receiver System(CMRS)/Infra-red Receiver System(IRRS)and was retired to MASDC on March 30, 1971, after 315 flights accumulating 615.4 flight hours. It was stored as FV027. (Lockheed Martin Tactical Aircraft Systems)

65-5709 (A1-27)
This test aircraft first flew on June 1, 1967, and delivered to the USAF on July 20, 1967. It was used at Nellis AFB for a short time in 1968 as part of the F-111D AIM-7G Sparrow missile flight test program. It was first retired to General Dynamics, Fort Worth on January 5, 1972. It had accumulated 206 flights and 539.3 flight hours. It was shipped to MASDC, now called AMARC, on February 12, 1990. (Lockheed Martin Tactical Aircraft Systems)

65-5710 (A1-28)
This test aircraft first flew on June 9, 1967 and was delivered to the USAF on August 8, 1967. 65-5710 was also used in the F-111D AIM-7G test program. It was retired at General Dynamics, Fort Worth on June 27, 1975 to be used for spare parts. It had accumulated 1,147.8 during 500 flights. During 1989, the fuselage was used for antenna testing. (Lockheed Martin Tactical Aircraft Systems)

66-0011 (A1-29)
66-0011 first flew on July 11, 1967, and was delivered to the USAF on July 31, 1967. In February 1968, it was assigned to the FB-111A flight test program replacing 65-5702. Photographed in September 1969, 66-0011 wears the Milky Way and Crest on its tactical camouflage. It was later assigned to the 4520th CCTS and used for Category III testing. It had FB-111 avionics installed and was used in testing for integration of the SRAM in the FB-111A. It was input to AMARC as FV004 on November 30, 1970, after flying a total of 203 missions and 473.0 flight hours. The aircraft has been scrapped. (Tom Brewer Collection)

66-0012 (A1-30)
The aircraft first flew on August 9, 1967, and was delivered to the USAF on August 19, 1967. This Harvest Reaper Jet was first assigned to the 4520th CCTS and was used as an AFFTC aircraft until October 30, 1969, when a failed engine inlet splitter plate failed and penetrated the fuselage fuel tank after landing. This caused a major fire. The aircraft was retired in place at Edwards AFB after flying 276 flights and accumulating 557.9 flight hours. It was trucked to Lowry AFB outside of Denver, Colorado, on April 20, 1972, to be used as a weapons loading trainer. (Craig Kaston Collection)

OPERATIONAL F-111As

One hundred twenty nine operational F-111As (factory numbers A1-31 through A1-159) were delivered between November 1967 and December 1968 with the majority going to the 474th Tactical Fighter Wing at Nellis AFB, Nevada. Most of the Fiscal Year 1966 aircraft spent their careers as training aircraft, assigned to the 442nd TFTS at Nellis and 389th TFTS at Mountain Home AFB, Idaho. While assigned to the 474th TFW, F-111A aircraft twice deployed to Thailand in support of the war in Southeast Asia. The first deployment, called Combat Lancer, took place between mid-March and the end of November 1968 and involved a total of eight aircraft. The second deployment, Constant Guard V/Linebacker, occurred in late September 1972 with two squadrons (429th TFS and 430th TFS) being deployed. The two squadrons of aircraft remained in Thailand until returning to Nellis in mid-June 1975. The F-111As remained at Nellis, assigned to the 474th TFW until July 1977, when they moved to Mountain Home AFB, Idaho, replacing the F-111Fs of the 366th TFW. The 366th TFW was the last assignment for the F-111A. All operational aircraft were either converted to EF-111As, sold to Australia, or retired to AMARC by June 1991.

66-0013 (A1-32)
The aircraft first flew on September 24, 1967, and was delivered to the USAF on October 16, 1967. It is seen here without unit markings at Nellis AFB during November 1967, shortly after delivery. This Harvest Reaper Jet was originally assigned to the 4527th Combat Crew Training Squadron (CCTS). On April 29, 1969, 0013 suffered a major fire as a result of an internal engine failure brought about by failure to install burner nozzle insulator supports during depot maintenance. The aircraft was repaired and returned to flying status. The aircraft was the 9th F-111A converted to an EF-111A. (Roy Lock)

66-0014 (A1-32)
Below: The aircraft first flew on October 13, 1967, and was delivered to the USAF on November 1, 1967. This Harvest Reaper Jet was originally assigned to the 4527th CCTS. Seen here on November 23, 1969, it wears the ND tail code and green and white chevron tail band of the 442nd TFTS (Tactical Fighter Training Squadron). The aircraft was the 12th F-111A converted to an EF-111A. (Roy Lock)

66-0015 (A1-33)
The aircraft first flew on October 22, 1967, and was delivered to the USAF on November 9, 1967. It's seen here on April 6, 1968, in the original light gray bottomed Southeast Asia Camouflage, without unit markings. This Harvest Reaper Jet was originally assigned to the 4527th CCTS. The aircraft was the 13th F-111A converted to an EF-111A. (Duane Kasulka via Gerry Markgraf)

CHAPTER THREE: U.S. AIR FORCE TACTICAL F-111s

66-0016 (A1-34)
66-0016 is seen here with the green tail stripe and crew name banner of the 442nd TFTS beneath the right crew hatch. The aircraft first flew on November 11, 1967, and was delivered to the USAF on November 27, 1967. It was modified to Harvest Reaper configuration and assigned to the 4480th TFS. 66-0016 took part in Combat Lancer, and, on March 18, 1968, flew the first Combat Lancer Mission attacking a truck park and storage area with 12 Mk 117 750 pound bombs. The aircraft was the 10th F-111A converted to an EF-111A. (Pat Martin Collection)

66-0017 (A1-35) NO PHOTO AVAILABLE
The aircraft first flew on November 16, 1967, and was delivered to the USAF on November 28, 1967. It was modified to Harvest Reaper configuration and assigned to 4481st TFS, 4480th TFW. It crashed and was destroyed on March 30, 1968 during a Combat Lancer mission. Its callsign was Hotrod 76. The crew, Major Sandy Marquardt and Capt Joe Hodges, successfully ejected and was recovered. The cause of the aircraft loss was believed to be foreign object (a tube of solidified sealant) which became lodged in the pitch/roll mixer assembly of the flight controls. This rendered the aircraft uncontrollable under certain conditions. The aircraft was on its 48th flight, having accumulated a total of 140.5 hours, when it crashed.

66-0018 (A1-36)
The aircraft first flew on November 12, 1967, and was delivered to the USAF on November 28, 1967. It was converted to Harvest Reaper configuration and assigned to the 4481th TFS. It deployed to Takhli on March 15, 1968, and took part in Combat Lancer, flying its first combat mission on March 25, 1968. It's seen here in March 1972, with the NA tail code and the 428th TFS Buccaneers emblem on the left fuselage side. This aircraft still has the green camouflage underside applied to the Harvest Reaper jets. The aircraft was the 11th F-111A converted to an EF-111A. (Author's Collection)

66-0019 (A1-37)
The aircraft first flew on December 8, 1967, and was delivered to the USAF on December 16, 1967. It was converted to Harvest Reaper configuration. It deployed to Takhli on March 15, 1968, and took part in Combat Lancer, during which a hydraulics problem gave it the distinction of being the first combat ground abort. Seen here, photographed on November 29, 1968, shortly after its return from Takhli, 66-0019 wears the Combat Lancer tail flash. The aircraft was the 6th F-111A converted to an EF-111A. (Roy Lock)

66-0020 (A1-38)
The aircraft first flew on December 17, 1967, and was delivered to the USAF on December 31, 1967. Though 60-0020 was modified to Harvest Reaper configuration. It did not deploy to Takhli as part of Combat Lancer. Seen here while assigned to the 57th FWW, 66-0020 wears the outline of the State of Nevada. The aircraft was the 5th F-111A converted to an EF-111A. (Author)

66-0021 (A1-39)
This aircraft first flew on December 18, 1967, and was delivered to the USAF on January 16, 1968. Photographed on June 31, 1981, 66-0021 wears the standard SEA camouflage with the black underside and the yellow tail stripe of the 389th TFTS. This Harvest Reaper aircraft took part in Combat Lancer. The aircraft was the 7th F-111A converted to an EF-111A. (Roy Lock)

CHAPTER THREE: U.S. AIR FORCE TACTICAL F-111s

66-0022 (A1-40)
Above: The aircraft first flew on December 22, 1967, and was delivered to the USAF on January 26, 1968, and modified to Harvest Reaper configuration. It is seen in this photo immediately prior to departure for its Combat Lancer deployment to Thailand. It was the first combat related F-111 loss, callsign Omaha 77. Since its manufacture it had flown 40 flights and accumulated 103.8 flight hours. The aircraft was destroyed March 28, 1968, two days after Combat Lancer operations started. Killed were Major Henry McCann and Captain Dennis Graham. The crash site was discovered during 1989, in the Phu Phan Mountain Range of Northeast Thailand. Crew members Identification "dog" tags were found with the wreckage. (Lockheed Martin Tactical Aircraft Systems)

66-0023 (A1-41)
Right: The aircraft first flew on January 17, 1968 and was delivered to the USAF on January 20, 1968. Photographed on November 29, 1968, 66-0023 wears the WF tail code and yellow and black checkerboard of the 4539th Fighter Weapons Squadron, 4525 Fighter Weapons Wing. The aircraft was a Harvest Reaper aircraft, but did not deploy to Takhli as part of Combat Lancer. It was the 17th F-111A converted to an EF-111A. (Bill Curry)

66-0024 (A1-42)
The aircraft first flew on January 4, 1968, and was delivered to the USAF on January 29, 1968. This Harvest Reaper aircraft, a Combat Lancer replacement aircraft, deployed to Takhli on April 1, 1968. This photo was taken immediately prior to its departure for Takhli. Flying as callsign Tailbone 78, it crashed and was destroyed on April 21, 1968, during Operation Combat Lancer, becoming the third and last F-111 lost during Combat Lancer. Neither the aircraft nor crew, (Commander David "Spade" Cooley USN, and Lt. Col. Ed Palmgren), was ever found. Some Combat Lancer crews believed the loss was due to failure of the horizontal stabilizer weld (same as the cause of the loss of 66-0032). Other crews believe that 0024 was flown into the ground as that crew thought they could fly lower at night using manual TFR than they could using auto TFR. The aircraft was on its 37th flight since it was manufactured and had accumulated 110.2 flight hours. (Lockheed Martin Tactical Aircraft Systems)

66-0025 (A1-43) NO PHOTO AVAILABLE
The aircraft first flew on January 13, 1968, and was delivered to the USAF on February 13, 1968. This Harvest Reaper jet was a Combat Lancer replacement aircraft. It deployed to Takhli for Combat Lancer on April 1, 1968. It crashed and was destroyed June 20, 1975, on the runway at Nellis AFB as a result of a catastrophic engine failure (caused by loss of a third stage compressor fan blade) while on a practice approach. The aircraft had flown 479 flights, accumulating 1,195.3 hours before the crash. This was the second loss during June 1975, of a Nellis F-111A due to engine failure. Nellis was operating with only one runway with scheduled repairs being accomplished on the primary runway. The 474th TFW Deputy Commander for Operations, Col. William Palmer, and Lt Col Robert Tidwell were the crew of 66-0025. The crash blew a large hole in the open runway, closing the runway and as a result, closing Nellis AFB to air traffic. The author was airborne as part of a four-ship flight just coming back from the Nellis bombing ranges and was forced to divert to Mountain Home AFB, Idaho, along with the other three F-111As in his flight.

66-0026 (A1-44)
The aircraft first flew on January 13, 1968, and was delivered to the USAF on February 14, 1968. Photographed on November 22, 1969, 66-0026 wears the ND tail code and green and white chevron tail band of the 442nd TFTS, and still has its original light gray underside. The aircraft was scheduled to be modified into an EF-111A. It crashed and was destroyed March 13, 1984 at Mountain Home AFB before EF-111A modifications could be accomplished. It had logged 1085 flights and accumulated 2,563.9 hours before the crash. Both crew members, Captains David Peth and Steven Locke, were killed. (Gerry Markgraf Collection)

66-0027 (A1-45)
Above: The aircraft first flew on January 13, 1968, and was delivered to the USAF on February 13, 1968. Photographed on August 24, 1980, 66-0027 wears the yellow tail stripe of the TFS, 366th TFW. The aircraft was the 8th F-111A converted to an EF-111A. (Ray Leader)

66-0028 (A1-46)
Right: The aircraft first flew on February 11, 1968, and was delivered to the USAF on March 20, 1968. 66-0028 was nicknamed *LA BOMBA* during 1976, wearing a white bomb with the name in black on both the left and the right nose gear doors. The aircraft was the 28th F-111A converted to an EF-111A. (Author)

CHAPTER THREE: U.S. AIR FORCE TACTICAL F-111s

66-0029 (A1-47)
This aircraft first flew on February 2, 1968, and was delivered to the USAF on March 31, 1968, and was a 474th TFW aircraft temporarily assigned to the 27th TFW at Cannon AFB. The aircraft crashed and was destroyed on September 1, 1971. It had accumulated 239 flights and 610.4 flight hours at the time of the crash. While flying at low level, the aircraft hit a turkey vulture which passed through the right windshield, with some fragments of the windshield hitting both the pilot and the WSO. As a result the crew became disoriented and lost control of the aircraft. The crew ejected successfully. Both crew members were injured by the bird remains and the fragments of glass from the destroyed windshield. (Lockheed Martin Tactical Aircraft Systems)

66-0030 (A1-48)
The aircraft first flew on February 2, 1968, and was delivered to the USAF on March 14, 1968. Photographed here on September 17, 1975, 66-0030 is in 442nd TFTS markings, flying a training mission from Nellis AFB. On March 24, 1978, the main landing gear A frame failed during taxi at Sacramento ALC. It was repaired and was the 36th F-111A converted to an EF-111A. (Author)

66-0031 (A1-49)
The aircraft first flew on February 12, 1968, and was delivered to the USAF on March 25, 1968. After having the safeing pins pulled from the SUU-20 practice bomb dispenser, 66-0031 taxis to the runway on September 17, 1975. The aircraft was the 4th F-111A converted to an EF-111A. (Author)

66-0032 (A1-50) NO PHOTO AVAILABLE
The aircraft first flew on February 12, 1968, and was delivered to the USAF on March 29,1968. It crashed and was destroyed on May 8, 1968, while assigned to the 474th TFW. The aircraft was almost brand new, having flown just 13 times, accumulating 27.9 flight hours. While flying auto TFR, medium ride at 1,000 feet AGL, and at approximately 480 knots, the aircraft pitched up followed by an immediate right roll. Aircraft control could not be regained, and the crew ejected. One crew member sustained a back injury. A failure in the horizontal tail actuator caused the right stabilator to drive full leading edge up, resulting in the loss of control of the aircraft.

66-0033 (A1-51)
The aircraft first flew on February 19, 1968, and was delivered to the USAF on April 15, 1968. Lit by the setting sun, 66-0033, as seen from the formation lead, heads north into the Nellis Range Complex during 1976. The aircraft was the 24th F-111A converted to an EF-111A. (Author)

66-0034 (A1-52)
Above: The aircraft first flew on March 1, 1968, and was delivered to the USAF on April 15, 1968. 66-0034, photographed here in 442nd TFTS markings on July 8, 1973, crashed and was destroyed on June 6, 1975, at Peach Springs, Arizona. It was on its 619th flight, having accumulated 1,585.8 flight hours. The aircraft suffered a fire following failure of the right engine fuel manifold. The manifold failure was probably caused by a failed engine fan blade penetrating the fuel manifold. The crew ejected safely. This was the first of three 474th F-111A losses due to engine failure, which occurred in 1975. The pilot, Captain Dave Bowles, was later killed in the third 1975 engine failure loss (66-0058) which occurred on October 7, 1975. (Bill Malerba via Tom Brewer)

66-0035 (A1-53)
Right: The aircraft was delivered to the USAF on April 19, 1968. 66-0035 was photographed in March 1976, in the Nellis AFB arming area. The aircraft was the 20th F-111A converted to an EF-111A. (Author)

CHAPTER THREE: U.S. AIR FORCE TACTICAL F-111s

66-0036 (A1-54)
The aircraft was delivered to the USAF on April 19, 1968. When photographed on October 2, 1971, 66-0036 is wearing a CC tail code and the red tail stripe of the 522nd TFS, 27th TFW. The aircraft was the 33rd F-111A converted to an EF-111A. (Bob Pickett)

66-0037 (A1-55)
The aircraft was delivered to the USAF on April 19, 1968. 66-0037 was photographed on September 18, 1980 in markings of the 389th TFS, 366th TFW. The aircraft was the 21st F-111A converted to an EF-111A. (Brian C. Rogers)

66-0038 (A1-56)
The aircraft first flew on March 25, 1968, and was delivered to the USAF on April 19, 1968. 66-0038 photographed on March 19, 1976 wears the green tail stripe of the 442nd TFTS. The aircraft was the 27th F-111A converted to an EF-111A. (Author)

66-0039 (A1-57)
The aircraft first flew on April 5, 1968, and was delivered to the USAF on July 31, 1968. When photographed in August 1979, 66-0039 wore the yellow tail stripe of the 389th TFS, 366th TFW. The aircraft was the 25th F-111A converted to an EF-111A. (Roy Lock)

66-0040 (A1-58) NO PHOTO AVAILABLE
The aircraft first flew on April 17, 1968, and was delivered to the USAF on August 1, 1968. While assigned to the 474th TFW, the aircraft crashed and was destroyed on September 23, 1968. The aircraft had flown just 16 times for a total of 35.6 flight hours when it crashed. Fuel imbalance, due to a fuel gage failure, caused an extreme aft center of gravity (cg) and as a result of the aft cg the aircraft lost control. The fuel gage in the F-111A actually controls the aircraft cg. The forward and aft tank needles on the gage have electrical contacts attached. When the fuel level in the forward tank lowers due to feeding fuel to the engines, the fuel pumps in the aft tank are turned on by the contacts on the gage needles, transferring fuel to the forward tank. With the gage failure, this automatic function did not occur, allowing the cg to move too far aft. The aircrew, U.S. Navy Lt. John Nash and RAAF Lt. Neal Pollock, ejected successfully.

66-0041 (A1-59)
The aircraft first flew on April 23, 1968, and was delivered to the USAF on August 17, 1968. Photographed at Eglin AFB on August 6, 1972 wears the markings of the 442nd TFTS. The aircraft was the second F-111A converted to an EF-111A. (Tom Brewer Collection)

66-0042 (A1-60)
The aircraft first flew on March 29, 1968, and was delivered to the USAF on April 30, 1968. 66-0042 was photographed on November 29, 1968, carrying the ND tail code of the 4527th CCTS. While assigned to the 474th TFW, the aircraft crashed and was destroyed on February 12, 1969. The aircraft had accumulated 60 flights and 238.1 flight hours. The aircraft crashed into the snow at 9,000 foot level of a 9,500 foot mountain near Elko, Nevada. Automatic TFR had been overridden by the pilot. Captains Robert Jobe and William Fuchlow, were killed. The wreckage, which was lost in the snow for over four months, was found on June 16, 1969. (Roy Lock)

66-0043 (A1-61) NO PHOTO AVAILABLE
The aircraft first flew on April 17, 1968, and was delivered to the USAF on July 24, 1968. The 474th TFW accumulated 60 flights and 238.1 flight hours. The aircraft was flying a chase aircraft for a Functional Check Flight (FCF) on March 4, 1969. While slowing to 220 knots without flaps or slats extended, the aircraft stalled. The crew misidentified the post stall gyration as a spin and implemented the wrong recovery procedure. The aircraft was then in a flight condition too low for a safe recovery and crashed near Texas Dry Lake, north of Nellis AFB. The crew ejected successfully.

66-0044 (A1-62)
The aircraft first flew on May 6, 1968, and was delivered to the USAF on July 31, 1968. 66-0044 is seen here returning to Nellis following a range mission in 1975. The aircraft is carrying a SUU-20 under each wing. This allowed the aircraft to fly 12 practice bomb runs on each mission. 66-0044 was the 14th F-111A converted to an EF-111A. (Author)

66-0045 (A1-63)
The aircraft first flew on May 8, 1968, and was delivered to the USAF on July 29, 1968. It is seen here in August 1980, with 389th TFS markings. While assigned to the 389th TFTS of the 366th TFW, 66-0045 suffered an inflight fire and loss of hydraulics on May 12, 1982. The fire was caused by an engine failure resulting from a fan blade failure. The aircraft was on its 1,232nd flight and had accumulated 3,007.8 flight hours when it crashed. The crew ejected successfully. The aircraft crashed and was destroyed at Saylor Creek Range, south of Mountain Home AFB. (Terry Love Collection)

66-0046 (A1-64)
The aircraft was delivered to the USAF on May 8, 1968. Seen here in September 1975, 66-0046 returns from the Nellis Range Complex with empty BRU-3A/A bomb rack. The aircraft was the 18th F-111A converted to an EF-111A. (Author)

66-0047 (A1-65)
The aircraft first flew on April 24, 1968, and was delivered to the USAF on August 10, 1968. 66-0047 is seen here in the markings of the 389th TFS, when photographed on April 26, 1980. The aircraft was the 15th F-111A converted to an EF-111A. (Brian C. Rogers)

66-0048 (A1-66)
The aircraft first flew on April 24, 1968, and was delivered to the USAF on August 16, 1968. Photographed on November 30, 1968, 66-0048 wears the WF tail code and yellow and black checkerboard of the 4539th Fighter Weapons Squadron, 4525 Fighter Weapons Wing, and still has the original light gray underside. The aircraft was the 29th F-111A converted to an EF-111A. (Author's Collection)

66-0049 (A1-67) NO PHOTO AVAILABLE
The aircraft first flew on June 13, 1968, and was delivered to the USAF on August 24, 1968. The aircraft was the first to be converted to the EF-111A configuration. It was initially an aerodynamic test bed with the canoe and antenna fairings added, but with no ECM equipment added.

66-0050 (A1-68)
Right: The aircraft first flew on June 14, 1968, and was delivered to the USAF on August 23, 1968. Photographed on November 3, 1979, 66-0050 carries the yellow tail stripe of the 389th TFS. The aircraft was the 30th F-111A converted to an EF-111A. (Don McGarry)

66-0051 (A1-69)
The aircraft first flew on June 13, 1968, and was delivered to the USAF on August 18, 1968. The aircraft was the third F-111A converted to an EF-111A. (Author)

66-0052 (A1-70)
The aircraft first flew on June 24, 1968, and was delivered to the USAF on July 21, 1968. 66-0052 was photographed on display at Offutt AFB, in mid July 1979, just two weeks before it was lost. The aircraft crashed and was destroyed on July 31, 1979, at Mountain Home AFB. The aircraft was on it 726th flight and had accumulated 1.883.5 hours when it crashed. While practicing close trail formation on a flight demonstration ride for a new pilot, 66-0052 flew through the lead aircraft's wake/engine exhaust, flaming out both engines. The aircraft entered a flat spin at 27,000 feet from which it did not recover. The crew did not attempt to eject. Captain Myles Hammon and 2 LT Larry McFarland were killed. (Author)

66-0053 (A1-71)
The aircraft first flew on June 16, 1968, and was delivered to the USAF on June 28, 1968. During 1969, 66-0053 was used for flight testing of the AN/AAR-34 and APS-109A/ALR-41 ECM receiver systems. It is seen here on February 17, 1972 at Eglin AFB in Tactical Air Warfare Center markings. The aircraft was used as the test bed for the Grumman/Norden Pave Mover battlefield surveillance radar, flying in this configuration during mid-1982. It was retired to AMARC, arriving on September 19, 1990, with a total of 1,039 flights and 2,020.5 flight hours. (Tom Brewer Collection)

66-0054 (A1-72)
The aircraft first flew on June 14, 1968, and was delivered to the USAF on August 15, 1968. 66-0054 is seen here in the markings of the 522nd TFS. It was delivered to Nellis AFB and assigned to the 474th TFW until moving to Mountain Home as part of operation Ready Switch. The aircraft while assigned to the 389th TFTS, 366th TFW, crashed and was destroyed on April 13, 1983, on the Saylor Creek Bombing Range near Mountain Home AFB. The aircraft departed controlled flight while recovering from high angle of attack maneuvers. The crew ejected successfully, but received back injuries. The aircraft had 1,266 flights and 3,104.3 flight hours when it crashed. (Author's Collection)

66-0055 (A1-73)
The aircraft first flew on June 25, 1968, and was delivered to the USAF on August 2, 1968. It is seen here at D-M during March 1982. The aircraft was the 37th F-111A converted to an EF-111A. (Gerald McMasters)

66-0056 (A1-74)
The aircraft first flew on June 6, 1968, and was delivered to the USAF on July 26, 1968. 66-0056 wears the WF tail code and yellow and black checkerboard of the 4539th Fighter Weapons Squadron, 4525 Fighter Weapons Wing, and still has the original light gray underside. The aircraft was the 23rd F-111A converted to an EF-111A. (Dave Menard Collection)

66-0057 (A1-75)
The aircraft was used as the structural integrity test backup aircraft and was the first F-111 to undergo Cold Proof Load Test. 66-0057 delivered to the USAF on August 8, 1969. It had been painted white in July, 1969. 66-0057 is seen here in its white paint at Edwards AFB, assigned to AFFTC, in September 1971. It was the 41st F-111A converted to an EF-111A. (Tom Brewer Collection)

66-0058 (A1-76)
The aircraft first flew on June 14, 1968, and was delivered to the USAF on September 6, 1968. Seen here, the aircraft was assigned to the 3246th Test Wing at Eglin AFB and flew with early small AD tail codes. It transferred to the 474th TFW in 1973. 66-0058 was flying on the wing of 67-0055 when 67-0055 collided with a Turbo-Commander on November 12, 1974. While assigned to the 474th TFW, callsign Tasty 15, 66-0058 crashed and was destroyed October 7, 1975, approximately ten miles north of the Indian Springs Auxiliary Airfield in Nellis range complex. Killed in the crash were Captain Dave Bowles, the pilot, and Major Merle "Don" Kenney, the WSO. The aircraft experienced a catastrophic engine failure and impacted a hill 30 feet from the top. Apparently the TFR had been deselected (paddled off), as the TFR, if left on, would have flown the aircraft over the hill. Captain Bowles had been the pilot of 66-0034 when it crashed on June 6, 1975. 66-0058 was the third 474th F-111A crash in 1975 caused by catastrophic engine failure. This crash, was the fourth F-111A catastrophic engine failure in five months and resulted in the grounding of the F-111As for almost four months. (Author's Collection)

67-0032 (A1-77)
The aircraft first flew on June 28, 1968, and was delivered to the USAF on August 4, 1968. In this photo, taken on March 3, 1979, 67-0032 wears the 389th TFS yellow tail stripe. The aircraft was the 40th F-111A converted to an EF-111A. (Ben Knowles)

67-0033 (A1-78)
The aircraft first flew on June 28, 1968, and was delivered to the USAF on July 29, 1968. In this photo taken in May 1982, 67-0033 wears the 389th TFTS yellow tail stripe. Black tail codes and tail numbers have replaced the earlier white markings. The aircraft was the 35th F-111A converted to an EF-111A. (Charles Trump)

67-0034 (A1-79)
The aircraft first flew on July 5, 1968, and was delivered to the USAF on August 18, 1978. In this photo taken on November 22, 1969, 67-0034 wears the NC tail code and red and white chevron of the 430th TFS Tigers. The aircraft was the 38th F-111A converted to an EF-111A. (Roy Lock)

67-0035 (A1-8)
The aircraft first flew on July 17, 1968, and was delivered to the USAF on October 29, 1968. In this photo, taken on May 19, 1979 67-0035 wears the 389th TFS yellow tail stripe. The aircraft was the 39th F-111A converted to an EF-111A. (Brian C. Rogers)

67-0036 (A1-81)
The aircraft first flew on July 7, 1958, and was delivered to the USAF on August 28, 1968. In this photo taken on December 12, 1969, 67-0036 wears the NC tail code and red and white chevron of the 430th TFS Tigers. The 474th TFW aircraft crashed and was destroyed on April 24, 1972, over the Grand Canyon while on its 221st flight and had accumulated 586.3 flight hours. 66-0036 departed controlled flight while doing rudder rolls. The pilot placed the flight control switch in the Takeoff & Land position. (Takeoff & Land position was designed to give the crew full rudder authority for low speeds used during takeoff and landing. It was not designed to be used in normal flight regimes.) The full rudder authority caused the aircraft to depart controlled flight and enter a right downward spiral. The crew could not recover the aircraft and ejected successfully. On landing, the capsule hit a canyon wall, rolled 55 feet downhill, coming to rest on its right side. When the parachute was released, it rolled an additional 64 feet, and came to rest inverted on its right side. Both crew members exited the capsule through the left hatch with only a few superficial scratches. (Tom Brewer Collection)

CHAPTER THREE: U.S. AIR FORCE TACTICAL F-111s

67-0037 (A1-82)
The aircraft first flew on July 10, 1968, and was delivered to the USAF on September 9, 1968. In this photo taken in March 1976, 67-0037 is in the arming area awaiting its turn to have the safeing pins removed from the practice bomb ejectors in the SUU-20s. The nose gear door is marked with skull and cross bones of the 428th Buccaneers of the 474th TFW. The aircraft was the 31st F-111A converted to an EF-111A. (Author)

67-0038 (A1-83)
The aircraft first flew on July 16, 1968, and was delivered to the USAF on August 31, 1968. 67-0038, seen here at Nellis in July 1976 carries the blue tail stripe of the 428th TFS and the red crew name banner of the 430th TFS. The aircraft was the 19th F-111A converted to an EF-111A. (Author)

67-0039 (A1-84)
The aircraft first flew on July 18, 1968, and was delivered to the USAF on October 24, 1968. 67-0039, flying off the wing of a KC-135A, carries the blue tail stripe of the 428th TFS. The aircraft was the 16th F-111A converted to an EF-111A. (Author)

67-0040 (A1-85) NO PHOTO AVAILABLE
The aircraft first flew on August 8, 1968, and was delivered to the USAF on October 28, 1968. The aircraft crashed and was destroyed on July 11, 1973, while assigned to the 474th TFW. The aircraft crashed in Zion National Park, five miles north of Springdale, Utah, following loss of the right windshield and both canopies as a result of a birdstrike. 67-0040 was on its 402nd flight and had accumulated 1,029.2 flight hours when it crashed. The crew, Major Robert M. Hopkins and Major Kirby Ludwig, ejected safely.

67-0041 (A1-86)
The aircraft was delivered to the USAF on October 30, 1968. Photographed at Nellis AFB on November 10, 1976, 67-0041 is marked with the blue tail stripe of the 428th Buccaneers. The aircraft was the 22nd F-111A converted to an EF-111A. (Author)

67-0042 (A1-87)
The aircraft was delivered to the USAF on October 28, 1968. 67-0042, carrying the skull and cross bones of the 428th TFS Buccaneers on the nose gear door, "takes the Nellis runway" as lead of a four ship formation. The aircraft was the 26th F-111A converted to an EF-111A. (Author)

67-0043 (A1-88) NO PHOTO AVAILABLE
The aircraft was delivered to the USAF on October 29, 1968. While assigned to the 474th TFW, 67-0043 crashed and was destroyed 50 miles northeast of Tuba City, Arizona on May 22, 1969, after a failure of the right windscreen. Speculation was a bird penetrated the windscreen, but no evidence of a bird strike was found on or in the capsule. The windshield failure caused the crew to become disoriented and lose control of the aircraft. The crew, Captain Kent May and Major John Morrow, ejected successfully. The evidence indicated a windshield failure, not a bird strike, and as a result emphasis was put on replacing the windshields with a stronger version.

67-0044 (A1-89)
The aircraft was delivered to the USAF on November 18, 1968. Seen here in the arming area at Nellis AFB on September 17, 1975, 67-0044 wears the blue tail stripe and the skull and cross bones of the 428th Buccaneers of the 474th TFW. The aircraft was the 42nd and last F-111A converted to an EF-111A. (Author)

67-0045 (A1-90)
The aircraft was delivered to the USAF on November 15, 1968. It's seen here on February 23, 1986, with the yellow tail stripe of the 389th TFTS, 366th TFW. 67-0045 was retired to AMARC on June 21, 1991, with a total of 4,569.8 flight hours retired. (Douglas Slowiak/Vortex Photo Graphics)

67-0046 (A1-91)
The aircraft was delivered to the USAF on November 15, 1968. Photographed at Norton AFB, California on October 21, 1983, 67-0046 was retired from the 389th TFTS, 366th TFW at Mountain Home, AFB on August 31, 1990, and had accumulated 1,952 flights and 4,627.4 flight hours. It was used as a ground trainer at Sheppard AFB. (Don McGarry)

67-0047 (A1-92)
The aircraft was delivered to the USAF on October 27, 1968. Photographed at Davis-Monathan AFB on January 25, 1983, 67-0047 was retired from the 389th TFTS, 366th TFW at Mountain Home, AFB on August 31, 1990, and had accumulated 2,031 flights and 4,855.1 flight hours. It was used as a ground trainer at Sheppard AFB. (Brian C. Rogers)

67-0048 (A1-93)
The aircraft was delivered to the USAF on November 22, 1968. Photographed in 1976, 67-0048 wears the blue tail stripe of the 428th TFS Buccaneers of the 474th TFW. The aircraft was the 34th F-111A converted to an EF-111A. (Author)

67-0049 (A1-94)
67-0049 was delivered to the USAF on November 26, 1968. As seen here photographed in November 1969, it crashed and was destroyed on December 22, 1969. The probable cause was a wing carry-through-box failure which occurred during a pull-up from a dive bomb run over the Nellis AFB Ranges. 67-0049 had 46 flights and 107.4 flight hours when it crashed. Ejecting out of the envelope, Lt Col Thomas Mack and Major James Anthony, were killed. This crash caused the USAF to begin the structural inspection and proof testing program for all F-111s. This inspection and testing program was called the Recovery Program. (Tom Brewer Collection)

67-0050 (A1-95)
The aircraft was delivered to the USAF on November 27, 1968. In January 1975, 67-0050 landed nose gear up at Nellis. The nose gear doors were open, but the nose gear failed to extend. The pilot, Lt Col Carl Hamby, lowered the nose into a path of foam applied to the center of the runway. The landing ground off most of the nose gear doors, blew the left main tire, and damaged the radome, but did not do major damage to the aircraft. 67-0050 arrived at AMARC from Sacramento ALC on July 28, 1994, with a total of 4,183.4 flight hours. (Author)

CHAPTER THREE: U.S. AIR FORCE TACTICAL F-111s

67-0051 (A1-96)
The aircraft was delivered to the USAF on November 19, 1968. It's seen here at Davis-Monathan AFB on January 25, 1983. 67-0051 was retired from the 389th TFTS, 366th TFW at Mountain Home, AFB on August 31, 1990, and had accumulated 2,019 flights and 4,751.4 flight hours. It was used as a ground trainer at Sheppard AFB. (Brian C. Rogers)

67-0052 (A1-97)
The aircraft was delivered to the USAF on August 15, 1969. It was photographed here in May 1979. The aircraft was the 32nd F-111A converted to an EF-111A. (Jim Goodall)

67-0053 (A1-98)
The aircraft was delivered to the USAF on November 20, 1968. In this photo taken in November 1969, 67-0053 wears the NA tail code and blue and white chevron of the 428th TFS Buccaneers of the 474th TFW. On May 4, 1973, the aircraft suffered major fire damage following an engine fire caused by an afterburner hydraulic pump failure. The aircraft was repaired and returned to flight status. 67-0053 arrived at AMARC on July 24, 1991, with a total of 2,032 flights and 4,631.2 flight hours. (Gerry Markgraf Collection)

67-0054 (A1-99)
The aircraft was delivered to the USAF on November 25, 1968. Photographed with the yellow tail stripe of the 389th TFTS, 67-0054 is seen here on the AMARC receiving line. The aircraft arrived at AMARC on June 19, 1991, with a total of 2,115 flights and 4,951.0 flight hours. (Douglas Slowiak/Vortex Photo Graphics)

67-0055 (A1-100) NO PHOTO AVAILABLE
The aircraft was delivered to the USAF on August 24, 1969. The aircraft was involved in a mid-air collision with a civilian Turbo-Commander, and crashed on November 12, 1974, near Kingston, Utah. 67-0055 as lead of a two ship formation (Sigma 71 and Sigma 72) was closing at approximately 320 knots for a night refueling with what the crew thought was KC-135A 58-0110, assigned to the 319th Bomb Wing at Grand Forks AFB, North Dakota. Instead of being a tanker, it was a civilian Turbo-Commander flying VFR (Visual Flight Rules) at 180 knots, in the refueling track. The F-111A collided with the Turbo-Commander destroying it and killing the pilot. Even though the F-111A had received damage to the cockpit area and the ejection capsule, the crew was able to successfully eject. 67-0055 had flown 441 flights and accumulated 1,113.5 hours at the time of the accident.

67-0056 (A1-101)
The aircraft was delivered to the USAF on August 29, 1969. 67-0056 was photographed here on April 16, 1984. The aircraft was retired from the 389th TFTS, of the 366th TFW at Mountain Home, AFB on August 31, 1990, and had accumulated 2,035 flights and 4,659.6 flight hours. It was used as a ground trainer at Sheppard AFB. (Douglas Slowiak/Vortex Photo Graphics)

67-0057 (A1-102)
The aircraft was delivered to the USAF on November 13, 1968. On November 6, 1969, while assigned to the 474th TFW, 67-0057 experienced a nacelle ejector duct failure. The aircraft was repaired and returned to flying status. Photographed here in April 1979, the aircraft wore the yellow tail stripe of the 389th TFS. 67-0057 was retired from the 366th TFW at Mountain Home AFB on August 31,1990, having accumulated 2,130 flights and 4,975.9 flight hours. It was used as a ground trainer at Sheppard AFB. (Author's Collection)

67-0058 (A1-103)
The aircraft was delivered to the USAF on January 20, 1969. The aircraft was a replacement aircraft for Linebacker II, arriving at Takhli on November 21, 1972.In this photo taken on March 16, 1984, 67-0058 wears markings as the flagship of the 366th TFW. It was retired from the 366th TFW on February 2, 1988, after flying 1,852 flights and accumulating 4,384.3 flight hours. 67-0058 was put on display on American Legion Blvd, Mountain Home, Idaho. (Douglas Slowiak/Vortex Photo Graphics)

CHAPTER THREE: U.S. AIR FORCE TACTICAL F-111s

67-0059 (A1-104)
The aircraft was delivered to the USAF on November 26, 1968. This 429th TFS Constant Guard V/Linebacker aircraft arrived in the second phase of the deployment. 67-0059 is seen here, photographed on December 7, 1974 at Korat RTAFB. The aircraft crashed and was destroyed on January 4, 1979, while assigned to the 388th TFTS, 366th TFW at Mountain Home AFB. The aircraft crashed after a catastrophic engine failure. The engine failure started a fire which eventually burned through the rudder controls and caused loss of control. The aircrew ejected successfully. 67-0059 had 940 flights accumulating 2,356.8 hours when it crashed. (Gerry Markgraf Collection)

67-0060 (A1-105)
Above: The aircraft was delivered to the USAF on November 26, 1968. In this photo, 67-0060 wears the NB tail code of the 429th TFS, 474th TFW. This 429th TFS Constant Guard V/Linebacker aircraft arrived in the first phase of the deployment. It crashed and was destroyed on April 7, 1976, near Wendover, Utah, while assigned to the 474th TFW. Following a left engine explosion and fire, aircraft control was lost. The crew ejected successfully with no injuries. 67-0060 had accumulated 750 flights and 1,910.6 flight hours at the time of the crash. (USAF)

67-0061 (A1-106)
Left: The aircraft was delivered to the USAF on December 31, 1968. This 429th TFS Constant Guard V/Linebacker aircraft arrived in the second phase of the deployment. Photographed here in 366th TFW markings, 67-0061 carries 24 MK 82 low drag 500 pound bombs. The aircraft arrived at AMARC on June 21, 1991, with a total of 5,029.4 flight hours. (USAF)

67-0062 (A1-107)
The aircraft was delivered to the USAF on December 17, 1968. Seen here at Nellis, 67-0062 is marked with a "111" above its tail code. This Constant Guard replacement arrived at Takhli on October 23, 1972. It was retired to AMARC on August 24, 1990, with total flight hours of 5387.4. (474th Roadrunners Association collection)

67-0063 (A1-108)
The aircraft was delivered to the USAF on January 9, 1969. This 429th TFS seen here was photographed on July 20, 1972. 67-0063 arrived in the second phase of the deployment. It was destroyed on November 6, 1972, in Southeast Asia, while flying with the callsign Whaler 57. It was the third Constant Guard V aircraft loss. The North Vietnamese Government claimed to have shot down this aircraft. The crew, Major Robert Brown and Captain Robert Morrissey, was not recovered. The North Vietnamese media reported the shoot down was accomplished by firing AAA tracers at the aircraft sound and instructed the air defense soldiers to fire their rifles and AK-47s in front of the tracers. The AAA trailed the aircraft and missed, but, as the North Vietnamese media reported, the F-111 flew into the barrage of small arms fire which brought it down. The media also claimed the same tactics were used in the shoot down of 67-0068, the last F-111 lost over North Vietnam. 67-0063 had 333 flights and had accumulated 822.1 flight hours when lost. (Dave Menard)

67-0064 (A1-109)
The aircraft was delivered to the USAF on December 31, 1968. Photographed here on March 21, 1971, 67-0064 wears the NB tail code and yellow tail stripe of the 429th TFS. This 429th TFS Constant Guard V/Linebacker aircraft arrived in the second phase of the deployment. 67-0064 was retired from the 366th TFW on June 30, 1988, with 1,957 flights and 4,567.7 flight hours. It was then used as a ground trainer for a short time, arriving at AMARC on August 28, 1990. (Author's Collection)

67-0065 (A1-110)
The aircraft was delivered to the USAF on January 17, 1969. This 429th TFS Constant Guard V/Linebacker aircraft arrived in the first phase of the deployment. In this photograph taken during the Constant Guard deployment, 67-0065 is carrying CBUs. The aircraft retired from the 366th TFW and arrived at AMARC on July 19, 1991, with a total of 5,522.9 flight hours. (USAF)

67-0066 (A1-111)
The aircraft was delivered to the USAF on February 6, 1969. This 429th TFS aircraft arrived in the second phase of the Constant Guard V/Linebacker deployment. It was destroyed on October 16, 1972 in Southeast Asia. 67-0066 is seen in this photo on the ramp at Takhli RTAFB callsign Coach 33. It was the second Constant Guard V aircraft lost, possibly hit by a SA-2 surface to air missile while flying near Phuc Yen Airfield. The belief was the aircraft was hit after climbing to MK84 GPLD delivery altitude which was above the normal combat TFR altitude. Neither the aircraft nor the crew, Captain James Hockridge and 1LT Allen Graham, was recovered. The Vietnamese government returned their remains on September 30, 1977. 67-0066 had 307 flights and accumulated 809.1 flight hours when lost. (Author's Collection)

67-0067 (A1-112)
The aircraft was delivered to the USAF on January 28, 1969. This Constant Guard V/Linebacker replacement aircraft arrived at Takhli on October 23, 1972. Seen here in 1976, 67-0067 carries the tiger and red crew banner of the 430th TFS Tigers and the blue tail stripe. After being assigned to the 474th TFW and the 366th TFW, it was retired to the USAF Museum, Wright-Patterson AFB Ohio. (Author)

67-0068 (A1-113)
The aircraft was delivered to the USAF on February 20, 1969. 67-0068 is seen here on takeoff from Nellis in August 1972, with 429th TFS NB tail codes and yellow tail stripe. The aircraft arrived in the first phase of the Constant Guard V/Linebacker II deployment. It was destroyed on December 22, 1972, in Southeast Asia. Callsign Jackal 33 was lost, probably hit by AAA, after bombing the Hanoi docks on the Red River. This was the sixth Constant Guard V aircraft lost. The crew, Captain Bob Sponeybarger, pilot, and 1Lt Bill Wilson, PWSO(Pilot-Weapon Systems Officer), ejected. They were the only F-111 crew confirmed held as POWs by the North Vietnamese government. They were returned to U.S. government control on March 29, 1973, as part of the formal release of POWs called for in the treaty ending U.S. involvement in the Vietnam War. 67-0068 had 372 flights and accumulated 946.8 flight hours when lost. (Mick Roth)

67-0069 (A1-114)
The aircraft was delivered to the USAF on February 26, 1969. This 429th TFS Constant Guard V/Linebacker aircraft arrived in the second phase of the deployment. 67-0069 is seen here with 366th TFW markings during October, 1983. It was retired from the 366th TFW and later used as a ground trainer at Indian Head, Maryland. (Terry Love Collection)

67-0070 (A1-115)
The aircraft was delivered to the USAF on March 11, 1969. This 429th TFS Constant Guard V/Linebacker aircraft arrived in the first phase of the deployment. After being assigned to the 391st TFS, 366th TFW, as seen here in September 1980, and later the 389th TFTS, 67-0070 was later used as a battle damage repair aircraft at McClellan AFB, California. (George Cockle)

67-0071 (A1-116)
The aircraft was delivered to the USAF on February 27, 1969. This 429th TFS Constant Guard V/Linebacker aircraft arrived in the second phase of the deployment. 67-0071 was involved in a mid-air collision with F-111A 67-0098 on February 17, 1973. The collision occurred during a formation position change on a pathfinder bombing mission. Both aircraft recovered to Udorn Royal Thai Air Force Base. They were repaired and returned to service. 67-0071 is seen here at Nellis during 1976 with crew banner and nose gear door markings of the 429th TFS Black Falcons. It was retired from the 366th TFW and arrived at AMARC on June 20, 1990, with a total of 2,327 flights and 5,420.2 flight hours. (Author)

67-0072 (A1-117)
The aircraft was delivered to the USAF on March 14, 1969. A 429th TFS Constant Guard V/Linebacker aircraft arrived in the first phase of the deployment. 67-0072 is seen in this photo at Takhli RTAFB on December 27, 1972. It crashed and was destroyed on February 20, 1973, at Takhli RTAFB, Thailand. The main landing gear pin failed during takeoff causing the aircraft to depart the end of the runway and burn. The crew egressed successfully. 67-0072 had accumulated 476 flights and 1,180.9 flight hours at the time of the loss. (Harley Copic)

67-0073 (A1-118)
Left: The aircraft was delivered to the USAF on March 7, 1969. This 430th TFS Constant Guard V/Linebacker aircraft arrived in the third phase of the deployment. It is seen here in November 1981 at Holloman AFB. While assigned to the 391st TFS, 366th TFW, it crashed and was destroyed on January 19, 1982, 11 miles northwest of Mountain Home AFB heading southwest. The crash was caused by an explosion and fire in the engine bay. This was caused by a failure of a patch on the forward fuselage (F-2) fuel tank which resulted in the spill of a massive amount of JP4 fuel into the right engine bay. The crew could not regain control after the rudder went "hardover" due to burn-through of the rudder controls. The aircrew ejected inverted, and at less than 2000 feet above the ground. The capsule did a slow roll after leaving the aircraft and landed upright. The WSO was briefly hospitalized for ejection spinal compression injuries. 67-0073 had accumulated 1,181 flights and 2,918.8 flight hours at the time of the crash. (Gerald McMasters)

67-0074 (A1-119)
The aircraft was delivered to the USAF on March 28, 1969. 67-0074 was photographed here on display for the George AFB, California Air Show in June 1971, wearing the NB tail codes and markings of the 429th TFS Black Falcons. This 429th TFS Constant Guard V/Linebacker aircraft arrived in the first phase of the deployment. 67-0074 was retired from the 366th TFW and arrived at AMARC on July 19, 1991, with a total of 2,269 flights and 5,157.6 flight hours. (Author)

67-0075 (A1-120)
The aircraft was delivered to the USAF on March 20, 1969. This 429th TFS Constant Guard V/Linebacker aircraft arrived in the first phase of the deployment. As seen in this photo taken in June 1991, 67-0075 was assigned to the 2874th Test Squadron at Sacramento-ALC, McClellan AFB. It was retired and arrived at AMARC on July 15, 1992, with a total of 4,775.6 flight hours. (Douglas Slowiak Collection)

67-0076 (A1-121)
The aircraft was delivered to the USAF on March 26, 1969. This 429th TFS Constant Guard V/Linebacker aircraft arrived in the second phase of the deployment. As seen it this photo, while assigned to the 429th TFS, during 1976 67-0076 was painted in a US Bicentennial paint scheme which was designed by the author while assigned to the 429th TFS. It was retired from the 366th TFW and arrived at AMARC on July 31, 1991, with a total of 2,333 flights and 5,409.6 flight hours. (Author)

67-0077 (A1-122)
The aircraft was delivered to the USAF on March 28, 1969. This 430th TFS Constant Guard V/Linebacker aircraft arrived in the third phase of the deployment. 67-0077 is seen with 390th TFS, 366th TFW markings, at Nellis in this photo taken on December 8, 1980. It was retired from the 366th TFW and arrived at AMARC on June 19, 1991, with a total of 2,405 flights and 5,441.5 flight hours. (Brian C. Rogers)

67-0078 (A1-123)
The aircraft was delivered to the USAF on April 7, 1969. Seen here on July 8, 1972, at Wright Patterson AFB, less than three months before deploying to Takhli as part of Constant Guard V. This 429th aircraft was the first Constant Guard V aircraft loss. Flying as Ranger 23, the aircraft crashed and was destroyed on September 28, 1972, during first night of bombing missions over North Vietnam. The aircraft was lost after hitting its target. No traces of either the aircraft or crew, Major William Coltman and 1LT Robert Brett Jr., were found. 67-0078 had accumulated 282 flights and 735.5 flight hours at the time of the crash. (Tom Brewer Collection)

67-0079 (A1-124)
The aircraft was delivered to the USAF on March 26, 1969. This 429th TFS Constant Guard V/Linebacker arrived in the first phase of the deployment. 67-0079 is seen here with the yellow tail stripe of the 429th TFS and carrying captive AIM-9 sidewinder missiles. The fairing on the right weapons bay door indicates that the aircraft is equipped with the weapons bay M61A1 20mm Gattling gun. The aircraft suffered a major fire on the ground after failure of a secondary air duct on January 1, 1981. 67-0079 was shipped to General Dynamics, Forth Worth on April 6, 1981. The engine inlet and lower fuselage were replaced at Fort Worth as part of the repair. The repair took over 13 months at a price of 2.5 million dollars. It returned to service on May 29, 1982. 67-0079 was retired from the 366th TFW and arrived at AMARC on September 18, 1990, with a total of 2,030 flights and 4,808.4 flight hours. (USAF)

67-0080 (A1-125)
The aircraft was delivered to the USAF on April 7, 1969. It's seen in this photo wearing markings of the 429th Black Falcons. This Constant Guard V/Linebacker aircraft arrived in the second phase of the deployment. While assigned to the 474th TFW at Nellis AFB, it crashed and was destroyed on March 11, 1976. Control of the aircraft was lost during an ECM formation, after an overshoot while attempting a rejoin on the lead aircraft. The crew ejected with both members sustaining back injuries. 67-0080 had accumulated 767 flights and 1,933.6 flight hours at the time of the crash. (USAF)

67-0081 (A1-126)
The aircraft was delivered to the USAF on April 18, 1969. This Constant Guard V/Linebacker replacement aircraft arrived in Southeast Asia on May 8, 1973. It is seen here at Nellis on July 12, 1975, after its return from Korat RTAFB still wearing the HG tail code of the 347th TFW. 67-0081 was retired from the 366th TFW and arrived at AMARC on June 21, 1991, with a total of 2207 flights and 4,992.2 flight hours. (Author)

67-0082 (A1-127)
The aircraft was delivered to the USAF on April 17, 1969. Seen here on February 19, 1972, 67-0082 wears the NB tail code and yellow tail stripe of the 429th TFS. The aircraft crashed and was destroyed shortly after takeoff on June 18, 1972, in Choctawtchee Bay, Florida, three miles from Eglin AFB. A fuel fire caused by an open gas cap on the top of the fuselage was believed to be the cause of the crash. The fuel ran back along the fuselage and was ignited by the afterburners. The fire caused loss of control. The ejection occurred at 5,484 feet above water. The parachute streamed and the module impacted the waster. Both crew members, Col Keith Brown and Lt Col James Black, sustained fatal injuries. 67-0082 had accumulated 237 flights and 609.4 flight hours at the time of the crash. (Tom Brewer)

67-0083 (A1-128)
The aircraft was delivered to the USAF on April 17, 1969. This 429th TFS Constant Guard V/Linebacker aircraft arrived in the first phase of the deployment. 67-0083 is seen here on November 10, 1976 at Nellis taxiing back from a training mission. It crashed and was destroyed on November 30, 1977, during a toss bomb run on the Nellis Ranges while on a Red Flag mission. Ejection was not attempted. Captain Art Stowe and Major Lorely "Skip" Wagner, were killed. 67-0083 had accumulated 816 flights and 2,095.1 flight hours at the time of the crash. (Author)

67-0084 (A1-129)
The aircraft was delivered to the USAF on April 21, 1969. 67-0084 was a 429th TFS Constant Guard V/Linebacker aircraft and arrived in the first phase of the deployment. Seen here on June 27, 1981, at McChord AFB, Washington, 67-0084 is marked with the MO tail code of 366th TFW. It was retired from the 366th TFW and arrived at AMARC on July 17, 1991, with a total of 2,369 flights and 5,501.8 flight hours. (Pat Martin)

67-0085 (A1-130)
The aircraft was delivered to the USAF on April 28, 1969. 67-0085 in this photo taken in October 1969, wears the NA tail code and blue and white chevron of the 428th TFS Buccaneers of the 474th TFW. As a 430th TFS Constant Guard V/Linebacker aircraft, 67-0085 arrived in the third phase of the deployment. It was retired from the 366th TFW and arrived at AMARC on June 21, 1990, with a total of 2,014 flights and 4,657.7 flight hours. (Gerry Markgraf Collection)

67-0086 (A1-131)
The aircraft was delivered to the USAF on April 29, 1969. This 429th TFS Constant Guard V/Linebacker aircraft was the first Constant Guard aircraft at Takhli. In this photo taken at Nellis AFB on February 14, 1975, 67-0086 prepares for takeoff with its wingman, an F-4E. 67-0086 was retired from the 366th TFW and arrived at AMARC on June 19, 1991, with a total of 2,458 flights and 5,438.9 flight hours. (Author)

67-0087 (A1-132)
The aircraft was delivered to the USAF on June 20, 1969. 67-0087 was assigned to the 391st TFS, 366th TFW when photographed in September 1986. It was assigned to the 429th TFS when deployed to Constant Guard V/Linebacker, arriving in the second phase of the deployment. 67-0087 was retired from the 366th TFW and arrived at AMARC on July 24, 1991, with a total of 2,551 flights and 5,799.5 flight hours. (Author's Collection)

CHAPTER THREE: U.S. AIR FORCE TACTICAL F-111s

67-0088 (A1-133)
The aircraft was delivered to the USAF on May 6, 1969. This 429th TFS Constant Guard V/Linebacker aircraft arrived in the second phase of the deployment. It is seen here in November 1975, at McClellan AFB, after returning from Korat RTAFB. 67-0088 was retired from the 366th TFW and arrived at AMARC on June 19, 1991, with a total of 2,271 flights and 5,203.0 flight hours. (Bill Strandberg)

67-0089 (A1-134)
The aircraft was delivered to the USAF on May 10, 1969. This 429th TFS Constant Guard V/Linebacker replacement aircraft arrived at Takhli on November 25, 1972. On August 14, 1975, 67-0089 was the third F-111A in three months to experience a #3 fan blade failure. The engine failure happened on the runway as the engines were being run up for takeoff. The aircraft had the wings removed and was shipped by C-5A to McClellan AFB for repair. A total of four F-111A aircraft suffered this failure between June 6, 1975, and October 7, 1975. 66-0024, 66-0034, and 66-0058 were lost due to this engine problem. Only 66-0058 was a fatal crash. 67-0089 seen here was photographed in December 1983 in markings of the 390th TFS, 366th TFW. It was retired from the 366th TFW with 1,537 flights and 3,536.6 flight hours, arriving at AMARC by truck on September 17, 1990. (Terry Love Collection)

67-0090 (A1-135)
The aircraft was delivered to the USAF on June 24, 1969. This 429th TFS Constant Guard V/Linebacker aircraft arrived in the second phase of the deployment. 67-0090 seen here was photographed on March 20, 1982 in markings of the 390th TFS, 366th TFW. It was retired from the 366th TFW and arrived at AMARC on July 19, 1991, with a total of 2,442 flights and 5,559.5 flight hours. (Douglas Slowiak/Vortex Photo Graphics)

67-0091 (A1-136)
Right: The aircraft was delivered to the USAF on May 20, 1969. Seen here in 430th TFS markings including a NC tail code in 1971, this 430th TFS Constant Guard V/Linebacker aircraft arrived in the third phase of the deployment. 67-0091 was retired from the 366th TFW and arrived at AMARC on July 31, 1991, with a total of 2,361 flights and 5,436.0 flight hours. (USAF)

67-0092 (A1-137)
The aircraft was delivered to the USAF on May 24, 1969. 67-0092 in this photo taken in October 1969, wears the NB tail code and yellow and white chevron of the 429th TFS Black Falcons of the 474th TFW. 67-0092 was assigned to the 430th TFS when deployed to Constant Guard V/Linebacker arriving in the third phase of the deployment. It was the fourth Constant Guard V loss. Flying with callsign Burger 54, it crashed and was destroyed on November 20, 1972 after going "feet wet" in the Gulf of Tonkin. Some wreckage was found, but Captains Ronald Stafford and Charles Cafferelli, were not recovered. The cause of the loss has not been determined/released. 67-0092 had accumulated 344 flights and 917.2 flight hours at the time of the loss. (Tom Brewer Collection)

67-0093 (A1-138)
The aircraft was delivered to the USAF on July 30, 1969. This 430th TFS Constant Guard V/Linebacker aircraft arrived in the third phase of the deployment. It is seen here in 390th TFS, 366th TFW markings when photographed on September 12, 1979. While assigned to the 390th TFS, it was destroyed in ground fire November 9, 1982, on the live ordnance ramp (called the SAC Pad) at Mountain Home AFB. The fire was fed by an oxygen leak in the cockpit. MK-82s on the aircraft exploded in a low order detonation and destroyed the aircraft. 67-0093 had accumulated 1,191 flights and 2,815.8 flight hours at the time of the loss. (Brian C. Rogers)

67-0094 (A1-139)
Above: The aircraft was delivered to the USAF on May 12, 1969. This 430th TFS, Constant Guard Aircraft arrived in the third phase of the deployment. On May 14, 1975, 67-0094, as part of the recovery of the captured U.S. merchant ship SS Mayaguez, dropped MK-48 GPLD (General Purpose Low Drag) bombs on Cambodian gunboats, sinking one. The 0094 survived a mid-air collision with 67-0111 on June 16, 1973. Both aircraft were accomplishing radar bomb runs against the same target in Cambodia. They were vectored together by airborne radar controllers. 67-0111 crashed. 67-0094 lost 4.5 feet of its wing and recovered to Ubon RTAFB, landing at 240 knots. 67-0094 was retired from the 366th TFW and arrived at AMARC on June 21, 1991, with a total of 2,494 flights and 5,622.7 flight hours. (Author)

Left: This gunboat kill marking was still on 67-0094 in August 1975, following its return from Korat RTAFB. (Author)

67-0095 (A1-140)
The aircraft was delivered to the USAF on May 29, 1969. This 430th TFS Constant Guard V/Linebacker aircraft arrived in the third phase of the deployment. 67-0095 seen here was photographed on February 27, 1988 in markings of the 366th TFW. It was retired from the 366th TFW and arrived at AMARC on July 24, 1991, with a total of 2,266 flights and 5,223.8 flight hours. (Brian C. Rogers)

67-0096 (A1-141)
The aircraft was delivered to the USAF on May 29, 1969. This 430th TFS Constant Guard V/Linebacker aircraft arrived in the third phase of the deployment. 67-0096 seen here was photographed on March 24, 1979 in markings of the 390th TFS, 366th TFW. It carries the Giant Voice -1979 emblem on the nose. 67-0096 was retired from the 366th TFW and arrived at AMARC on August 24, 1990, with a total of 2,420 flights and 5,498.6 flight hours. (Brian C. Rogers)

67-0097 (A1-142)
The aircraft was delivered to the USAF on May 27, 1969. This 430th TFS Constant Guard V/Linebacker aircraft arrived in the third phase of the deployment. As seen here, when photographed on September 12, 1979, 67-0097 was assigned to the 390th TFS, 366th TFW, when it crashed and was destroyed on March 26, 1980, on Saylor Creek Range, south of Mountain Home AFB. The aircraft departed controlled flight during training. Captains Joseph G. Raker and Larry Honza, were killed after ejecting below the safe ejection envelope. 67-0097 had accumulated 1097 flights and 2,718.8 flight hours at the time of the crash. (Brian C. Rogers)

67-0098 (A1-143)
Right: The aircraft was delivered to the USAF on June 11, 1969. This 430th TFS Constant Guard V/Linebacker arrived in the third phase of the deployment. 67-0098 was involved in a mid-air collision with F-111A 67-0071 on February 17, 1973. The collision occurred during a formation position change on a pathfinder bombing mission. Both aircraft recovered to Udorn Royal Thai Air Force Base. They were repaired and returned to service. It is seen here on May 13, 1979 in markings of the 390th TFS, 366th TFW. 67-0098 crashed and was destroyed on October 8, 1982, while assigned to the 390th TFS, 366th TFW, Mountain Home AFB. It was on its second flight after having gone through depot maintenance. The AC electric power leads had not been tightened properly during the depot maintenance. They loosened causing flight control problems. The crew ejected successfully, but the impact attenuation bags failed, severely injuring the spinal cord of the pilot, Wing Commander Col. Ernest Coleman. The WSO was lucky and did not receive major injuries. 67-0098 had accumulated 1,002 flights and 2,315.2 flight hours at the time of the crash. (Bob Pickett)

67-0099 (A1-144) NO PHOTO AVAILABLE
The aircraft was delivered to the USAF on June 11, 1969. This 430th TFS aircraft arrived in the third phase of the deployment. It was the fifth Constant Guard V loss, flying as Snub 40. It crashed and was destroyed December 18, 1972, after going "feet wet" in the Gulf of Tonkin. Neither the aircraft nor the crew, Lt Col Ronald Ward and Major James McElvain, was found. The cause of the loss could not be determined. 67-0099 had accumulated 336 flights and 853.7 flight hours at the time of the loss.

67-0100 (A1-145)
The aircraft was delivered to the USAF on June 17, 1969. This 430th TFS Constant Guard V/Linebacker aircraft arrived in the third phase of the deployment. Seen here at Nellis on December 8, 1987, 67-0100 is marked as the *Spirit of Idaho*, the wing commander's aircraft. It was retired from the 366th TFW in July 1991, with 2,511 flights and 5,680.3 flight hours. It is now on display in the aircraft park at Nellis AFB. (Craig Kaston Collection)

67-0101 (A1-146)
The aircraft was delivered to the USAF on June 16, 1969. This 430th TFS Constant Guard V/Linebacker aircraft arrived in the third phase of the deployment. 67-0101 seen here was photographed on July 11, 1981 in markings of the 390th TFS, 366th TFW. On August 2, 1982, it suffered a major inflight fire due to an engine ingesting a bird. It diverted into Idaho Falls, Idaho, and was later shipped to General Dynamics, Fort Worth for repair. It returned to service in August 1982. 67-0101 was retired from the 366th TFW and arrived at AMARC on August 24, 1990, with a total of 1,852 flights and 4,304.0 flight hours. (Brian C. Rogers)

67-0102 (A1-147)
The aircraft was delivered to the USAF on June 20, 1969. 67-0102 is seen here in 430th TFS markings including a NC tail code on February 19, 1972. This 430th TFS Constant Guard V/Linebacker aircraft arrived in the third phase of the deployment. The aircraft crashed and was destroyed on January 12, 1988, while assigned to 366th TFW at Mountain Home AFB. On takeoff from Mountain Home, the right crew hatch popped opened. The pilot reduced speed, but the aircraft was not configured for the lower speed (flaps and slats had already been retracted). It departed controlled flight, and the aircrew ejected while inverted. Captains Robert Meyer and Frederick Gerhart were killed when the capsule impacted the ground. 67-0102 had accumulated 1,618 flights and 3,719.7 flight hours at the time of the loss. (Tom Brewer Collection)

67-0103 (A1-148)
This 430th TFS Constant Guard V/Linebacker aircraft arrived in the third phase of the deployment. 67-0103 seen here was photographed on September 12, 1983, in markings of the 390th TFS, 366th TFW. It was retired from the 366th TFW on April 30, 1990, with 1,935 flights and 4,384.3 flight hours. It was used as a GF-111A ground trainer until arriving at AMARC on January 30, 1993. (Brian C. Rogers)

67-0104 (A1-149)
The aircraft was delivered to the USAF on June 26, 1969. This 430th TFS Constant Guard V/Linebacker aircraft arrived in the third phase of the deployment. 67-0104 seen here was photographed in April 1979, in markings of the 390th TFS, 366th TFW. It was retired from the 366th TFW and arrived at AMARC on August 24, 1990, with a total of 2,366 flights and 5,505.1 flight hours. (Terry Love)

67-0105 (A1-150)
The aircraft was delivered to the USAF on June 27, 1969. 67-0105 in this photo wears the NB tail code and yellow and white chevron of the 429th TFS Black Falcons of the 474th TFW. While assigned to the 430th TFS, this aircraft deployed to Constant Guard V/Linebacker arriving in the third phase of the deployment. It crashed and was destroyed on July 5, 1979, while assigned to the 390th TFS, 366th TFW. The aircraft crashed when flying as lead of a two ship formation during a daylight mission on the Nellis ranges. Major Gary Mekash and Lt Col Eugene Soeder were killed. 67-0105 had accumulated 893 flights and 2,200.5 flight hours at the time of the loss. (Dave Menard Collection)

67-0106 (A1-151)
The aircraft was delivered to the USAF on July 10, 1969. This 430th TFS Constant Guard V/Linebacker arrived in the third phase of the deployment. The red tail stripe and red crew name banner on this photo indicate it was assigned to the 430th TFS, 474th TFW, when photographed on February of 1976. 67-0106 was retired from the 366th TFW and arrived at AMARC on July 31, 1991, with a total of 2,338 flights and 5,263.9 flight hours. (Author)

67-0107 (A1-152)
The aircraft was delivered to the USAF on July 10, 1969. This 430th TFS Constant Guard V/Linebacker aircraft arrived in the third phase of the deployment. Seen here, photographed in May 1986, 67-0107 wears the yellow tail stripe of the 389th TFTS, 366th TFW. It was retired from the 366th TFW and arrived at AMARC on July 17, 1991, with a total of 2,459 flights and 5,600.4 flight hours. (Wayne Whited)

67-0108 (A1-153)
The aircraft was delivered to the USAF on July 17, 1969. 67-0108 in this photo taken in November 1969, wears the NB tail code and yellow and white chevron of the 429th TFS Black Falcons of the 474th TFW. This aircraft was a Constant Guard V replacement aircraft arriving at Takhli on January 1, 1973. 67-0094 was retired from the 366th TFW and arrived at AMARC on July 24, 1991, with a total of 2,599 flights and 5,892.5 flight hours. (Gerry Markgraf Collection)

67-0109 (A1-154)
The aircraft was delivered to the USAF on July 15, 1969. This was a Constant Guard V replacement aircraft arriving at Takhli on December 11, 1972. 67-0109 seen here was photographed on July 11, 1981 with the green tail stripe of the 390th TFS, 366th TFW. It was purchased by RAAF in June 1982. 67-0109 was modified to F-111C configuration and given the RAAF number A8-109. It is presently assigned to the No. 82 Wing, RAAF Amberley. (Mike Grove)

67-0110 (A1-155)
The aircraft was delivered to the USAF on July 29, 1969. This 430th TFS Constant Guard V/Linebacker aircraft arrived in the third phase of the deployment. 67-0110 seen here was photographed on February 26, 1979 with 366th Wing markings for Giant Voice – the 1979 SAC Bombing Competition. It was retired from the 366th TFW and arrived at AMARC on August 24, 1990, with a total of 2,325 flights and 5,337.1 flight hours. (Brian C. Rogers)

67-0111 (A1-156)
The aircraft was delivered to the USAF on July 18, 1969. 67-0111 in this photo taken on November 23, 1969, at a Nellis Open House wears the NB tail code and yellow and white chevron of the 429th TFS Black Falcons of the 474th TFW. This Constant Guard V/Linebacker aircraft replacement aircraft arrived in Southeast Asia on May 8, 1973. The seventh, and last combat loss under the Constant Guard V/Linebacker deployment, it crashed and was destroyed in Southeast Asia (Cambodia) on June 16, 1973. The crash was a result of a mid-air collision with 67-0094 on June 16, 1973. Both aircraft were accomplishing radar bomb runs against the same target in Cambodia, and were vectored together by airborne radar controllers. The crew of 67-0111 ejected and was recovered. 67-0094 lost 4.5 feet of its wing, landing at Ubon RTAFB. 67-0111 had accumulated 332 flights and 836.2 flight hours at the time of the loss. (Roy Lock)

67-0112 (A1-157)
The aircraft was delivered to the USAF on July 22, 1969. This 430th TFS Constant Guard V/Linebacker aircraft arrived in the third phase of the deployment. 67-0112 is seen here, photographed with the ND tail code of the 442nd TFTS. It was later assigned to the 430th TFS, 474th TFW, and, after Operation Ready Switch, to the 391st TFS, 366th TFW. It was transferred from the 366th TFW and purchased by RAAF in June 1982. 67-0112 was modified to F-111C configuration and given the RAAF number A8-112. It is presently assigned to the No. 82 Wing, RAAF Amberley. (C. Roger Cripliver Collection)

67-0113 (A1-158)
The aircraft was delivered to the USAF on August 14, 1969. This 430th TFS Constant Guard V/Linebacker aircraft arrived in the third phase of the deployment. This aircraft flew the last combat mission of the war in Southeast Asia, on August 15, 1973, against a target in Cambodia. While assigned to Mountain Home, the aircraft landed 600 feet short of the runway and was out of commission for more than two years. 67-0113 seen here was photographed in December 1980 in markings of the 390th TFS, 366th TFW. The aircraft was purchased by RAAF in June 1982, modified to F-111C status, and numbered A8-113. It is presently assigned to the No. 82 Wing, RAAF Amberley. (Terry Love Collection)

67-0114 (A1-159)
The aircraft was delivered to the USAF on August 6, 1969. This 430th TFS Constant Guard V/Linebacker aircraft arrived in the third phase of the deployment. 67-0114 seen here was photographed in May 1979 in markings of the 390th TFS, 366th TFW. 67-0114 was retired from the 366th TFW. The aircraft was purchased by RAAF in June 1982, modified to F-111C status, and numbered A8-114. It is presently assigned to the No. 82 Wing, RAAF Amberley. (Douglas Slowiak/Vortex Photo Graphics)

EF-111A

In the late 1960s and early 1970s, the Air Force was looking for a replacement for its Douglas EB-66 electronic warfare aircraft. Although the EB-66 still continued to fly combat missions over North Vietnam, its engines were unreliable and the aircraft was rapidly beginning to show signs of age.

Because of the small number of aircraft required by the U.S. Air Force, the development of a completely new type of aircraft for this role was not economically practical. The Grumman EA-6B Prowler carrier-based electronic warfare aircraft was considered, but the Air Force was reluctant to acquire an aircraft originally developed for the Navy. The Air Force concluded in 1972, that modification of existing F-111A would be a cost-effective solution. This decision resulted in the EF-111A.

The EF-111A would be equipped with a version of the AN/ALQ-99 noise-jamming system used on the EA-6B. The Air Force calculated the EF-111A would have an unrefueled on-station loiter time of eight hours compared with two and one half hours for the Navy/Grumman EA-6B. This added endurance would make the EF-111A available for successive strikes. The EF-111A's Mach 2.2 speed gave the aircraft improved survivability. The EA-6B system designed to be operated by three Electronic Warfare Officers (EWO), had to be modified to be operated by just one in the EF-111. As a result, the Tactical Jamming System (TJS) of the EA-6B was modified to a single EWO configuration.

In December 1974, Grumman was selected as the prime contractor for the conversion. On January 30, 1975, Grumman was awarded a contract for the modification of two F-111As (66-0049 and 66-0041) as EF-111A prototypes.

The modifications included the installation of an AN/ALQ-99E jamming subsystem. Exciters, antennae, and other items most were mounted on a pallet inside the internal weapons bay. Other components were mounted inside a 16-foot ventral canoe shaped radome. A fin-tip pod accommodated the electronic countermeasures receivers. The self-protection subsystem consisted of a jamming system and a countermeasures dispensing set. A terminal threat warning subsystem was installed which consisted of infrared and electronic countermeasures receiver sets. The vertical fin had to be reinforced in order to support the fin-tip pod, new electrical wiring had to be installed, 60 kVA generators were replaced by 90 kVA units, and an improved environmental system for electronic equipment cooling was installed. The cockpit had to be rearranged to accommodate the new Tactical Jamming System controls, with the flight controls being removed from the right side of the cockpit, and the navigation equipment being relocated so that the pilot in the left seat could use it. The controls and displays for the EWO were installed in the right side of the cockpit.

These modifications resulted in an increase of empty weight from 46,172 pounds for the F-111A to 55,275 pounds for the EF-111A. The lack of weapon carriage requirements resulted in its maximum takeoff weight being only 88,848 pounds, less than the 98,850 pounds of the F-111A. The EF-111A initially retained the Pratt & Whitney TF30-P-3 turbofans.

In December 1975, Grumman first flew a partially-modified F-111A fitted with a mockup of the ventral canoe located where the weapons bay doors had been. The first EF-111A prototype (serial number 66-0049) flew from the Grumman plant at Calverton, Long Island on March 10, 1977. It was in full external configuration with fin-tip pod and ventral fairing, but did not have most of its electronic equipment installed. The airframes to be converted were selected from F-111As A-31 through A-97 (66-0013 through 67-0052) which had been primarily used by the 474th TFW for aircrew training. The first flight of a fully-equipped EF-111A (serial number 66-0041) was made on May 17, 1977. This aircraft was delivered to Detachment 3 of the Tactical Air Warfare Center and stationed

at Mountain Home AFB for operational test and evaluation. The system performed satisfactorily, and six more aircraft were programmed to be modified, followed by 34 more, for a total of 42 EF-111A aircraft.

The name Raven was officially adopted for the EF-111A. In November 1981, the first deliveries of operational EF-111As were made to the 388th Electronic Combat Squadron (ECS) and later to the 390th ECS, both squadrons of the 366th Tactical Fighter Wing at Mountain Home AFB, Idaho. EF-111As were also delivered to the 42nd ECS of the 20th TFW at RAF Upper Heyford in February 1984. By December 1985, all 42 of the Ravens had been delivered to the 366th TFW and the 20th TFW, with Mountain Home AFB receiving 29 and Upper Heyford receiving 13. One of Mountain Home's EF-111As (66-0013) was assigned to Detachment 3 of the Tactical Air Warfare Center (TAWC).

The 66th Electronic Combat Wing (ECW) was based at Sembach AB, West Germany. On July 1, 1985, control of the 42nd (ECS) flying EF-111As at RAF Upper Heyford, England, transferred from the 20th TFW to the 66th ECW. The EF-111A Ravens remained at Upper Heyford, and continued to wear the UH tail codes of the 20th TFW. On January 25, 1991, after returning from deployments to the Gulf the 42nd ECS was again assigned to the 20th TFW. The 66th ECW was inactivated on March 31, 1992. On October 1, 1994 it was activated as 66th Air Base Wing, a non-flying unit.

The first combat mission for the Raven took place during Operation El Dorado Canyon, the retaliatory attack on Libya on the night of April 14-15, 1986. During that mission, the 42nd ECS provided three EF-111As plus two spare aircraft to jam the Libyan radar network. (See El Dorado Canyon in the "Combat Operations" chapter).

On April 14 and 15, 1986, aircraft of the 42nd ECS took part in Operation El Dorado Canyon. During the Gulf War, aircraft from the 42nd ECS were assigned to the 7440th Wing (Provisional) at Incirlik, Turkey from late 1990, with others being a part of the 48th TFW (Provisional) at Taif.

In 1986, General Dynamics installed more powerful TF30-P-9 turbofans in the EF-111A. In January 1987, Grumman and TRW, Inc. started development of the Avionics Modernization Program (AMP) kit for the EF-111A. This kit provided the EF-111A with improved terrain following and navigational radars, a ring laser gyro inertial navigation system, the global positioning system receiver, two digital computers, improved cockpit displays, and upgraded communication systems. The first AMP kit was installed in EF-111A 66-0018 in January 1989. The majority of the EF-111As received this upgrade.

In support of combat actions in Iraq, five Upper Heyford EF-111As were deployed to Incirlik, Turkey as part of Operation Proven Force, and Eighteen EF-111A Ravens were deployed to Taif, Saudi Arabia in support of Operation Desert Storm. The Ravens flew over 900 sorties. None were lost in combat, but one was lost in a non-combat related accident and both crew members were killed. An EF-111A Raven, (66-0016), even claimed a "kill" during Desert Storm. On the night

of January 17, 1991, an Iraqi Mirage F.1 flew into the ground while chasing EF-111A 66-0016. Even though the Raven was unarmed with no air-to-air weapon capability, the Raven crew claimed the kill, backed by the Wing, but the kill was never credited to the crew. (See Desert Shield/Desert Storm in the "Combat Operations" chapter).

With the drawdown of F-111s in the USAF, the EF-111As moved from Upper Heyford to Cannon AFB. A detachment of EF-111As from the 429th ECS had been based at Incirlik AB, Turkey, since 1991. They returned home to Cannon AFB on June 24, 1997, after having flown more than 2,800 sorties in support of Operations Provide Comfort, Northern Watch, and

Southern Watch. The first EF-111As were retired in July and August 1997. 67-0041 was the first one to be retired arriving at AMARC on July 30, 1997. Some EF-111As remained at Cannon AFB until May 1998, when the last EF-111A was formally retired from the Air Force inventory, closing out the USAF career of the Aardvark.

EF-111A Deliveries

1	66-0049 (Prototype)	-	March 10, 1977 to Mountain Home AFB on November 20, 1981
2	66-0041	-	May 15, 1977 to Upper Heyford AB on May 23, 1984 after testing was complete
3	66-0051 (1st TAC)	-	November 11, 1981 to Mountain Home AFB
4	66-0031	-	May 28, 1982 to Mountain Home AFB
5	66-0020	-	April 15, 1982 to Mountain Home AFB
6	66-0019	-	May 21, 1982 to Mountain Home AFB
7	66-0021	-	June 25, 1982 to Mountain Home AFB
8	66-0027	-	September 16, 1982 to Mountain Home AFB
9	66-0013	-	November 30, 1982 to Mountain Home AFB assigned to TAWC Det 3
10	66-0016	-	December 14, 1982 to Mountain Home AFB
11	66-0018	-	March 24, 1983 to Mountain Home AFB
12	66-0014	-	May 12, 1983 to Mountain Home AFB
13	66-0015	-	June 6, 1983 to Mountain Home AFB
14	66-0044	-	July 28, 1983 to Mountain Home AFB
15	66-0047	-	July 22, 1983 to Mountain Home AFB
16	67-0039	-	August 17, 1983 to Mountain Home AFB
17	66-0023	-	October 7, 1983 to Mountain Home AFB
18	66-0046	-	October 28, 1983 to Mountain Home AFB
19	67-0038	-	November 28, 1983 to Mountain Home AFB
20	66-0035	-	December 9, 1983 to Mountain Home AFB
21	66-0037 (1st USAFE)	-	January 30, 1984 to Upper Heyford AB
22	66-0041	-	March 5, 1984 to Upper Heyford AB
23	66-0056	-	April 4, 1984 to Upper Heyford AB
24	66-0033	-	May 7, 1984 to Upper Heyford AB
25	66-0039	-	June 28, 1984 to Upper Heyford AB
26	67-0042	-	July 6, 1984 to Mountain Home AFB
27	66-0038	-	August 30, 1984 to Mountain Home AFB
28	66-0028	-	September 10, 1984 to Mountain Home AFB
29	66-0048	-	October 1, 1984 to Mountain Home AFB
30	66-0050	-	November 9, 1984 to Mountain Home AFB
31	67-0037	-	November 26, 1984 to Mountain Home AFB
32	67-0052	-	December 18, 1984 to Upper Heyford AB
33	66-0036	-	January 30, 1985 to Mountain Home AFB
34	67-0048	-	February 27, 1985 to Mountain Home AFB
35	67-0033	-	March 25, 1985 to Mountain Home AFB
36	66-0030	-	May 2, 1985 to Upper Heyford AB
37	66-0055	-	September 13, 1985 to Upper Heyford AB
38	67-0034	-	June 28, 1985 to Upper Heyford AB
39	67-0035	-	August 12, 1985 to Upper Heyford AB
40	67-0032	-	September 12, 1985 to Upper Heyford AB
41	66-0057	-	October 8, 1985 to Upper Heyford AB
42	67-0044	-	December 23, 1985 to Mountain Home AFB

66-0041 was placed in Grumman's Anechoic chamber for testing of the jamming system. (Paul Hart Collection)

EF-111A Raven

Forty-two EF-111As were modified from F-111As, A1-31 through A1-97, assigned to the 366th TFW at Mountain Home AFB at the time of modification. Initially 29 returned to the 366th TFW at Mountain Home AFB, and 13 were delivered to the 20th TFW at Upper Heyford. The EF-111As moved from Mountain Home AFB to Cannon AFB during 1992 and 1993, and from Upper Heyford to Cannon AFB during mid-1992. The EF-111As were retired from Cannon AFB to AMARC during 1997 and 1998.

66-0013 (A1-31) (EF-09)
Below: 66-0013 was the ninth aircraft modified into an EF-111A and was primarily used for operational testing flying from Mountain Home AFB while assigned to Det 3, 4485th Test Squadron, of the Tactical Air Warfare Center (TAWC) at Eglin AFB. In this photo, taken on April 15, 1990, 66-0013 carries OT tail code and a modified tail number indicating its assignment to DET 3. It was last assigned to the 429th ECS, 27th FW at Cannon AFB, New Mexico. (Chris Mayer)

66-0014 (A1-32) (EF-12)
66-0014, the 12th aircraft modified, is seen here in December 1988 in the markings of the 390th ECS, 366th TFW at Mountain Home AFB. It deployed to Saudi Arabia from the 390th ECS at Mountain Home AFB and took part in Operation Desert Storm. It was last assigned to the 429th ECS, 27th FW at Cannon AFB, New Mexico. (Author's Collection)

66-0015 (A1-33) (EF-13)
The 13th aircraft modified was delivered to Mountain Home on June 6, 1983. It was last assigned to the 429th ECS, 27th FW at Cannon AFB, New Mexico. It's seen here in May 1995, with CC tail code of the 27th FW and the 429th ECS Black Falcon on its tail. (Alec Fushi)

66-0016 (A1-34) (EF-10)
Right: A veteran of Combat Lancer, 66-0016 was the first F-111 ever to fly a combat mission, and was the 10th F-111A modified to EF-111A configuration. 66-0016 is seen here in June 1986 with the MO tail code of the 366th TFW. It deployed to Saudi Arabia from the 42nd ECS at Upper Heyford and took part in Operation Desert Storm. The aircraft claimed an Iraqi Mirage F1 kill on January 17, 1991, after flying at an altitude which caused the chasing Mirage to impact the ground. It was last assigned to the 429th ECS, 27th FW at Cannon AFB, New Mexico. (Terry Love)

66-0018 (A1-36) (EF-11)
This former Harvest Reaper/Combat Lancer jet was delivered to Mountain Home AFB, the 11th EF-111A modified, on March 24, 1983. 66-0018 was photographed with markings and MO tail code of the 366th TFW. It was last assigned to the 429th ECS, 27th FW at Cannon AFB, New Mexico. (Brian C. Rogers)

66-0019 (A1-37) (EF-06)
This former Harvest Reaper/Combat Lancer jet, the sixth modified, was delivered to Mountain Home AFB as an EF-111A on May 21, 1982. 66-0019 is seen here in May 1995, as the 429th ECS commanders aircraft. It was last assigned to the 429th ECS, 27th FW at Cannon AFB, New Mexico. (Alec Fushi)

66-0020 (A1-38) (EF-05)
This former Harvest Reaper jet was delivered to Mountain Home AFB as the fifth EF-111A on April 15, 1982. 66-0020 is seen here on June 11, 1989, with the MO tail code of the 366th TFW. It was last assigned to the 429th ECS, 27th FW at Cannon AFB, New Mexico. (Brian C. Rogers)

66-0021 (A1-39) (EF-07)
This former Harvest Reaper/Combat Lancer jet was delivered to Mountain Home AFB as the seventh aircraft modified to an EF-111A on June 25, 1982. It is seen here in 27th FW Wing flagship markings at the USAF 50th Anniversary Air Show at Nellis AFB in April 1997. It was last assigned to the 429th ECS, 27th FW at Cannon AFB, New Mexico. (Author)

66-0023 (A1-41) (EF-17)
Right: This former Harvest Reaper jet was the 17th modified to an EF-111A and was delivered to Mountain Home AFB on October 7, 1983. 66-0023 was photographed in May 1985 in the markings of the 366th TFW. It deployed to Saudi Arabia from the 42nd ECS at Upper Heyford and took part in Operation Desert Storm. While flying as Wrench 08 on February 14, 1991, the aircraft crashed and was destroyed in Saudi Arabia. Captains Douglas Bradt and Paul Eichenlaub II, were killed. (Author's Collection)

66-0027 (A1-45) (EF-08)
The aircraft was delivered to Mountain Home AFB as an EF-111A on September 16, 1982 as the eighth aircraft modified to an EF-111A. 66-0027 is seen here in markings of the 390th ECS, 366th TFW. It deployed to Saudi Arabia from the 390th ECS at Mountain Home AFB, and took part in Operation Desert Storm. It was last assigned to the 429th ECS, 27th FW at Cannon AFB, New Mexico. (Douglas Slowiak/Vortex Photo Graphics)

66-0028 (A1-46) (EF-28)
66-0028, the 28th EF-111A modified, was delivered to Mountain Home AFB as an EF-111A on September 10, 1984. 66-0028 is seen here in 390th ECS markings on a December 14, 1989, Red Flag mission. It was photographed from the copilots seat of the B-52 which it was escorting. It was last assigned to the 429th ECS, 27th FW at Cannon AFB, New Mexico. (Brian C. Rogers)

66-0030 (A1-48) (EF-36)
Right: The aircraft was delivered to Upper Heyford as the 36th EF-111A modified on May 28, 1985. 66-0030 was one of the six EF-111As which took part in Operation El Dorado Canyon, the attack on Libya. The aircraft deployed to Saudi Arabia from the 390th ECS at Mountain Home AFB and took part in Operation Desert Storm. It is seen here near the parking shelters at Taif, Saudi Arabia. It was nicknamed *Mild And Bitter Homebrew* between 1987 and 1988. It was last assigned to the 429th ECS, 27th FW at Cannon AFB, New Mexico. (USAF)

66-0031 (A1-49) (EF-04)
Below: 66-0031, the fourth modified, was first delivered to Mountain Home AFB as an EF-111A on May 28, 1982, and is seen in this photo on October 12, 1991. It was last assigned to the 429th ECS, 27th FW at Cannon AFB, New Mexico. The aircraft retired to AMARC on August 8, 1997. It flew to AMARC using the callsign Harpo 73 – the callsign used by the EF-111s during Operation El Dorado Canyon. (Keith Snyder)

66-0033 (A1-51) (EF-24)
The aircraft was delivered to Upper Heyford as the 24th EF-111A on May 7, 1984. 66-0033 was one of the six EF-111As which took part in Operation El Dorado Canyon attack on Libya. The aircraft deployed to Saudi Arabia from the 390th ECS at Mountain Home AFB and took part in Operation Desert Storm. It was nicknamed *Excalibur* between 1987 and 1988. Photographed in March 1996, 66-0033 shows the tail number presentation 6033 used at Cannon to identify the aircraft as 66-0033, so as not to be confused with 67-0033, also assigned to the 27th FW. It was last assigned to the 429th ECS, 27th FW at Cannon AFB, New Mexico. It was retired to AMARC on August 8, 1997. (Author's Collection)

CHAPTER THREE: U.S. AIR FORCE TACTICAL F-111s

66-0035 (A1-53) (EF-20)
66-0035 was first delivered to Mountain Home AFB as the 20th EF-111A on December 9, 1983. Photographed on August 4, 1996, it is marked at the DET 7 flagship with OT tail codes. It was last assigned to the 429th ECS, 27th FW at Cannon AFB, New Mexico and took part in Operation Proven Force. The aircraft was retired to AMARC on September 16, 1997. (Andrew H. Cline)

66-0036 (A1-54) (EF-33)
66-0036 was first delivered to Mountain Home AFB as the 33rd EF-111A on January 30, 1985. It's seen here in May 1995, in the markings of the 429th ECS at Cannon AFB, its last assignment (Alec Fushi)

66-0037 (A1-55) (EF-21)
The aircraft was delivered to Upper Heyford on January 30, 1984. The 21st aircraft modified was the first EF-111A assigned to USAFE. It was named *NATO Raven One* on February 2, and renamed *Prowler* during 1987-1988. The aircraft deployed to Saudi Arabia from the 390th ECS at Mountain Home AFB and took part in Operation Desert Storm. It is seen here in September 1993, assigned to the 429th ECS, 27th FW at Cannon AFB, New Mexico, its last assignment. (Dave Brown)

66-0038 (A1-56) (EF-27)
Right: 66-0038, the 27th modified, was delivered to Mountain Home AFB on August 30, 1984. 66-0038 was flown to General Dynamics, Fort Worth on November 6, 1985, for testing of the Jamming Sub-System update. When the program was canceled, the aircraft returned to Mountain Home on April 8, 1988. It's seen here on July 25, 1988 with MO tail codes of the 366th TFW. 66-0038 deployed to Saudi Arabia from the 42nd ECS at Upper Heyford and took part in Operation Desert Storm. It suffered a catastrophic failure to the wing carry through box during a Cold Proof Load Test (ground testing) in December 1991 at SM-ALC, McClellan AFB. 66-0038 was retired to AMARC on October 3, 1997. (Gerry Markgraf Collection)

66-0039 (A1-57) (EF-25)
The 25th aircraft modified, it was delivered to Upper Heyford as an EF-111A on June 28, 1984. It was named *The Sorcerer's Apprentice* from 1987-1988. 66-0039 is seen here in markings of the 429th ECS. It was last assigned to the 429th ECS, 27th FW at Cannon AFB, New Mexico and took part in Operation Proven Force. (Douglas Slowiak/Vortex Photo Graphics)

66-0041 (A1-59) (EF-02)
66-0041 was the second F-111A modified to EF-111A and was the first to have all the EF equipment installed. It flew system trials at Grumman beginning in June 1975, flying with all systems installed on May 15, 1977. 66-0041 is seen here in May 1979, with the bright red-orange safety markings used for ease of visibility during flight test. After completing testing, it was brought up to production standards and delivered to the 20th TFW at Upper Heyford on May 5, 1984. It was last assigned to the 429th ECS, 27th FW at Cannon AFB, New Mexico. The aircraft retired to AMARC on July 31, 1997. It flew to AMARC using the callsign Harpo 71, the callsign used by the EF-111s during Operation El Dorado Canyon. The aircraft retired to AMARC on September 16, 1997. (Author's Collection)

66-0044 (A1-62) (EF-14)
Originally not planned to be modified to EF-111A, until after 66-026 crashed, 66-0044 was delivered, as the 14th aircraft modified, to Mountain Home AFB as an EF-111A on July 28, 1983. The aircraft deployed to Saudi Arabia from the 390th ECS at Mountain Home AFB and took part in Operation Desert Storm. 66-0044 is seen here on April 4, 1994 at Langley AFB, in the markings of the 429th ECS, 27th FW. The aircraft crashed and was destroyed June 17, 1996, near Cannon AFB, New Mexico (Tucumcari). The crew ejected safely. (Brian C. Rogers)

66-0046 (A1-64) (EF-18)
The 18th aircraft modified, it was delivered to Mountain Home AFB as an EF-111A on October 28, 1983. 66-0046 deployed to Saudi Arabia from the 390th ECS at Mountain Home AFB and took part in Operation Desert Storm. It is seen here at a Red Flag at Nellis AFB on April 19, 1990. It was last assigned to the 429th ECS, 27th FW at Cannon AFB, New Mexico. (Brian C. Rogers)

66-0047 (A1-65) (EF-15)
The aircraft was delivered to Mountain Home AFB as the 15th EF-111A on July 22, 1983. 66-0047 is seen here in June 1984, with MO tail codes of the 366th TFW. It deployed to Incirlik, Turkey as part of Operation Proven Force. The aircraft was used as a testbed for the System Improvement Program (SIP), an upgrade to the aircraft's jamming subsystem. From 1993-1996 it was stationed at Eglin AFB where it underwent extensive flight testing to validate the equipment. Although the upgrades significantly enhanced the aircraft's jamming capabilities, the program was scrapped when the announcement came that the EF-111As were being retired. The last time it flew was in June 1996, from Eglin AFB to Cannon AFB. It is now used exclusively as a trainer for maintenance personnel. (Author's Collection)

66-0048 (A1-66) (EF-29)
Above: The aircraft was delivered to Mountain Home AFB as the 29th EF-111A on October 1, 1984. 66-0048 is seen here October 1, 1997, at Cannon AFB, assigned to the 429th ECS, 27th FW. It took part in Operation Proven Force. (Jerry Geer)

66-0049 (A1-67) (EF-01)
Right: The aircraft was the first EF-111A prototype. It's first test flight occured on March 10, 1977. It was initially used as the aerodynamic testbed without an operational system installed. It was reworked and brought up to production standards. 66-0049 is seen here in flight test paint scheme in June 1978. It was delivered to the 366th TFW at Mountain Home AFB on November 20, 1981, and took part in Desert Storm. It was last assigned to the 429th ECS, 27th FW at Cannon AFB, New Mexico. (Jim Rotramel)

66-0050 (A1-68) (EF-30)
The aircraft was delivered to Mountain Home AFB as the 30th EF-111A on November 9, 1984. 66-0050 is seen here on June 29, 1985, in 366th TFW markings. It deployed to Saudi Arabia from the 42nd ECS at Upper Heyford, and took part in Operation Desert Storm. It was last assigned to the 429th ECS, 27th FW at Cannon AFB, New Mexico. (Doug Remington)

CHAPTER THREE: U.S. AIR FORCE TACTICAL F-111s

66-0051 (A1-69) (EF-03)
The third aircraft modified was delivered to Mountain Home AFB as Tactical Air Command's (TAC's) first operational EF-111A arriving on November 5, 1981. 66-0051, seen here in October 1993, was last assigned to the 429th ECS, 27th FW at Cannon AFB, New Mexico. (Ben Knowles Collection)

66-0055 (A1-73) (EF-37)
The 37th aircraft modified was delivered to Upper Heyford as an EF-111A on September 13, 1985. It was named *Boomerang* from 1987-1988, and took part in Desert Storm. Seen here, 66-0055 has special 20th TFW tail markings outlined in white. It deployed to Incirlik, Turkey as part of Operation Proven Force. It was last assigned to the 429th ECS, 27th FW at Cannon AFB, New Mexico. (Author's Collection)

66-0056 (A1-74) (EF-23)
The aircraft was delivered to Upper Heyford as the 23rd EF-111A on April 3, 1984. It was named *Babyjam* from 1987-1988. 66-0056 is seen here on June 18, 1984, in the markings of the 42nd ECS, 20th TFW. The aircraft deployed to Saudi Arabia from the 42nd ECS at Upper Heyford and took part in Operation Desert Storm. The aircraft crashed and was destroyed April 2, 1992 at Finmere, UK due to a fuel duct failure. It carried the name *Jam Master* when lost. (Scott Wilson)

66-0057 (A1-75) (EF-41)
The aircraft was delivered to Upper Heyford as the 41st EF-111A on October 8, 1985. It was named *Special Delivery* from 1987-1988. The aircraft was one of the six EF-111As which took part in Operation El Dorado Canyon attack on Libya. It's seen here in markings of the 42nd ECS, 20th TFW. The aircraft deployed to Saudi Arabia from the 390th ECS at Mountain Home AFB, and took part in Operation Desert Storm. It was last assigned to the 429th ECS, 27th FW at Cannon AFB, New Mexico. (USAF)

67-0032 (A1-77) (EF-40)
Above: The 40th aircraft modified was delivered to Upper Heyford as an EF-111A on September 12, 1985. It was named *Black Sheep* from 1987-1988. Seen here on May 27, 1989, 67-0032 wears the white outlined tail markings of the 42nd ECS, 20th TFW. It was last assigned to the 429th ECS, 27th FW at Cannon AFB, New Mexico. (Tom Kaminski Collection)

67-0033 (A1-78) (EF-35)
Right: The aircraft was delivered to Mountain Home AFB as the 35th EF-111A on March 25, 1985. 67-0033 seen here on September 5, 1988, in the markings of the 390th ECS, 366th TFW was last assigned to the 429th ECS, 27th FW at Cannon AFB, New Mexico. (Douglas Slowiak/Vortex Photo Graphics)

67-0034 (A1-79) (EF-38)
Left: The 38th aircraft modified was delivered to Upper Heyford as an EF-111A on September 13,1985. It was named *Let 'Em Eat Crow* from 1987-1988. 67-0034 is seen here marked as the 42nd ECS flagship. The tail number marked on this aircraft looks like it is 66-0042; however, 734 indicating its true serial number is marked on the nose gear door. 67-0034 was one of the six EF-111As which took part in Operation El Dorado Canyon attack on Libya. The aircraft deployed to Incirlik, Turkey as part of Operation Proven Force. It was last assigned to the 429th ECS, 27th FW at Cannon AFB, New Mexico. (Author's Collection)

67-0035 (A1-80) (EF-39)
Below: The aircraft was delivered to Upper Heyford AB as the 39th EF-111A on August 12, 1985. On December 16, 1985, 67-0035 was damaged by fire and later returned to flying status. It was named *Ye Olde Crow* from 1987-1988, as seen in this photo. It was last assigned to the 429th ECS, 27th FW at Cannon AFB, New Mexico. (Author's Collection)

67-0037 (A1-81) (EF-31)
The aircraft was delivered to Mountain Home AFB as the 31st EF-111A on November 26, 1984. 67-0037 deployed to Saudi Arabia from the 390th ECS at Mountain Home AFB, and took part in Operation Desert Storm. It is seen in this photo, taken at Nellis AFB on April 19, 1990, in 390th ECS, 366th TFW markings. It was last assigned to the 429th ECS, 27th FW at Cannon AFB, New Mexico. (Brian C. Rogers)

67-0038 (A1-82) (EF-19)
Right: The aircraft was delivered to Mountain Home AFB as the 19th EF-111A on November 28, 1983. 67-0038 seen here in 366th TFW markings deployed to Saudi Arabia from the 390th ECS at Mountain Home AFB and took part in Operation Desert Storm. It was last assigned to the 429th ECS, 27th FW at Cannon AFB, New Mexico. (Mike Grove)

67-0039 (A1-83) (EF-16)
Below: The 16th aircraft modified was delivered to Mountain Home AFB as an EF-111A on August 17, 1983. The aircraft, seen here in markings of the 390th ECS, deployed to Saudi Arabia from Mountain Home AFB and took part in Operation Desert Storm. It was last assigned to the 429th ECS, 27th FW at Cannon AFB, New Mexico. (Author's Collection)

67-0041 (A1-86) (EF-22)
The aircraft was delivered to Upper Heyford as the 22nd EF-111A on March 5, 1984. It was named *Night Jammer* from 1987-1988. 67-0041 was one of the six EF-111As which took part in Operation El Dorado Canyon attack on Libya. The aircraft deployed to Saudi Arabia from the 390th ECS at Mountain Home AFB and took part in Operation Desert Storm. As seen here in May 1995, 67-0041 was last assigned to the 429th ECS, 27th FW at Cannon AFB, New Mexico, taking part in Operation Proven Force. It was the first EF-111A to be retired, arriving at AMARC on July 30, 1997. (Alec Fushi)

CHAPTER THREE: U.S. AIR FORCE TACTICAL F-111s

67-0042 (A1-87) (EF-26)
The 26th aircraft modified was delivered to Mountain Home AFB as an EF-111A on July 6, 1984. It was named *Spirit of Idaho* as one of two 366th TFW Flagships. The aircraft deployed to Incirlik, Turkey as part of Operation Proven Force. It is seen here in 390th ECS markings with 32 Desert Storm mission markings. It was last assigned to the 429th ECS, 27th FW at Cannon AFB, New Mexico. (Jim Geer)

67-0044 (A1-89) (EF-42)
The aircraft was the last EF-111A delivered, arriving at Mountain Home AFB on December 23, 1985. As seen here in this photo taken in July 1987, 67-0044 was last assigned to the 429th ECS, 27th FW at Cannon AFB, New Mexico. (Jim Goodall)

67-0048 (A1-93) (EF-34)
The aircraft was delivered to Mountain Home AFB as the 34th EF-111A on February 27, 1985. 67-0048 served with Det 3 of TAWC during the late 1980s. As seen in this photo taken on October 1, 1997, 67-0048 was last assigned to the 429th ECS, 27th FW at Cannon AFB, New Mexico. (Jerry Geer)

67-0052 (A1-97) (EF-32)
The aircraft was delivered to Upper Heyford as 32nd EF-111A on December 18, 1984. It was named *Cherry Bomb* from 1987-1988. 67-0052 was one of the six EF-111As which took part in Operation El Dorado Canyon attack on Libya. As seen in this photo taken on May 18, 1990, 67-0052 was assigned to the 429th ECS, 366th at Mountain Home AFB. It was last assigned to the 27th FW at Cannon AFB, New Mexico. (Pat Martin)

F-111D

Though out of sequence alphabetically, the F-111D's predecessor was the F-111E. The F-111D had a number of new features which included Mark II avionics system (sometimes referred to as the Mark IIA Avionics Subsystem), upgraded environmental control system, and TF30-P-9 engines.

The decision was made by Secretary McNamara to begin contract definition on Mark II avionics systems for both the strategic (FB-111A) and tactical F-111s in accordance with Air Force Advanced Development Objective (ADO No. 53) of March 1964. This ADO reflected a November 1963, recommendation for an improved avionics system (Mark II) to control in any weather the release of various air-to-air missiles against high and low altitude targets.

The future Mark II-equipped F-111A was designated F-111D one year before endorsement of the earlier F-111E. The "D" got underway on May 10, 1967, when the contract for a total of 493 F-111s replaced the basic production contract of April 1965. A concurrent System Management Directive (SMD) specified the Mark II avionics system would be installed in 132 F-111s, starting with the 236th production aircraft. In June 1966 an advanced contract change notified General Dynamics of this requirement. The F-111B, F-111C, F-111K, and FB-111A aircraft were counted in the 493 production aircraft under contract, but not in the USAF tactical production sequence.

On May 26, 1967, another USAF SMD gave the Mark II-equipped F-111D the radar-controlled AIM-7G-1 (Sparrow) air-to-air missile. The requirement for this missile was in addition to an improved, infrared, heat-seeking, air-to-air missile, similar to that of the F-111A and the F-111E. The Hughes AIM-4D (Falcon) and the Philco-Raytheon AIM-9D (Navy Sidewinder), were considered, but dropped in favor of the familiar Philco-General Electric AIM-9B (Sidewinder IA) of the F-111A, F-111E, and many other USAF fighters. After the June 23, 1966, Mark II contract award to Autonetics (a division of the North American Rockwell Corporation), a request for adaptation of the new Raytheon-developed YAIM-7G Sparrow to the Mark II's fire-control radar was added.

The May 1967, acquisition program gave the future Mark II-equipped F-111D the P-3 engines of the basic aircraft. Successful efforts to devise a more reliable and higher-thrust engine for the FB-111A interim bomber changed this planning. The Air Force decided in mid-1968 that the future F-111D would be equipped with the P-9, another version of the Pratt and Whitney TF30 turbofan. The new engine, first flight tested using an F-111A on July 10, 1968, entered production in early 1969. The P-9 featured the small afterburner of the P-1 and P-3 engines for greater thrust, the nozzle of the FB-111A's P-7 for more efficient thrust control, and the fan and low-pressure compressor of the Navy F-111B's P-12 for operating at higher engine temperatures. The P-9's thrust surpassed the P-3's, but was still well below the engine thrust the Air Force desired for the F-111D.

In early 1968, it was expected the Mark II avionics would add $1.5 million to the cost of each F-111D. This estimate was quickly revised upward to $2.2 million. By mid-1972 actual RDT&E costs of each F-111D already exceeded $4 million. These cost increases in the Mark II system with a tight budget reduced the F-111D program buy to enough aircraft to supply only one Combat Wing. The Air Force disclosed on September 12, 1969, it would limit Mark II production to 96 aircraft. The 96 F-111Ds would equip the 27th TFW's four squadrons (481st TFTS, 522nd TFS, and 524th TFS, along with the 4427th TFRS) with 18 aircraft each, leaving 24 F-111Ds for testing, replacement, and support. The balance of F-111Ds under procurement would receive a cheaper avionics package and would be known as F-111Fs.

The Air Force decided in December 1969 to put FB-111A tires on the F-111D's main and nose landing gears. F-111D main landing gear's axles, axle pins, stabilizer rods, as well as attachment pins and nuts, would also be replaced with FB-111A hardware. F-111Ds already off the production line (but not released for lack of Mark II avionic systems) were retrofitted, as were all F-111A and F-111E aircraft. The F-111F also benefited with the engineering changes being incorporated into the first production F-111F. This allowed the new aircraft to carry more fuel and a heavier weapon load. In March 1970, development of the Raytheon AIM-7G Sparrow was canceled leaving the future F-111D armed like other tactical F-111s with the capability to carry six air-to-air AIM-9B Sidewinders and one 20mm M61A1 Gatling gun (mounted on the right inside of the weapon bay).

Since the test program was primarily geared to test the aircraft's new avionics, the whole test program slipped until the avionics were ready for testing. The Category I tests which had been set to start in October 1967 did not start until December 1968. In September 1970, almost two years later, additional Category I flight tests were authorized to evaluate the Mark II's Integrated Display Set (IDS) in a new production configuration. Development problems slipped the delivery of a first and incomplete prototype of the Mark II system until June 1968. General Dynamics flew the prototype avionics for the first time on December 2, 1968, 14 months late, installed in an F-111A.

By late 1969, a complete Mark II avionics system was still not available, and the system's escalating cost had reduced the F-111D program to 96 aircraft, instead of 315 originally planned for production. Redesigns, engineering changes, additional requirements, and the like accounted for the cost overruns. But the economy-dictated F-111D reduction caused component costs to swell as mass production slumped.

The Air Force further intended to use an F-111A to begin Category II testing. However, the mid-1968 decision to give the F-111D a new engine with the Triple Plow II air diverter changed this planning. The Air Force dedicated five early F-111D production aircraft for testing. The first flight of an F-111D, the first F-111D production (68-0085), occurred on May

15, 1970, at the General Dynamics' Fort Worth plant, with the aircraft accepted by the USAF on June 30, 1970. This occurred the day after the lifting of the six-month F-111 delivery hold-order, which was imposed after the F-111A crash of December 22, 1969. This aircraft had undergone most of the cold-proof, structural test Recovery Program required after that F-111A loss. The first F-111D was equipped with the new P-9 engine, but without a complete Mark II system.

After accepting one F-111D in June 1970, no more were delivered for the next 12 months. The unavailability of Mark II avionics systems was the reason for the delay. The F-111Ds were not exempted from the recovery program and were produced on a schedule independent of the Mark II's availability. By late 1970, General Dynamics had completed most of the F-111D airframes, the last 50 receiving the Recovery inspections during production. Lacking an avionics system, a first increment of 40 airframes was parked at the Fort Worth plant in mid-1970, awaiting the outcome of a new round of Mark II contractual and production arrangements.

Despite every effort, F-111D deliveries proceeded slowly when they resumed in July 1971. Only 24 of 96 F-111Ds were available in June 1972. This was two years after the 27th Tactical Fighter Wing should have been operational. The 27th TFW, Cannon AFB, received F-111Es beginning in September 1969. These aircraft went to USAFE's 20th TFW one year later, but there were no F-111Ds to replace them at Cannon, and the operational ready status was not reached until mid-1973.

The F-111D finally entered operational service on November 1, 1971. The F-111D first saw service with the 27th TFW at Cannon. The first aircraft programmed for operational service (the sixth F-111D produced – 68-0090) was accepted by the Air Force on October 28, 1971. Its first flight occurred on September 28, 1971. It was equipped with a full Mark II avionics system. Initial Operational Capability (IOC) occurred during September 1972, by one of the 27th wing's three tactical fighter squadrons, three years later than originally planned.

The 27th TFW increased its monthly average strength of F-111Ds from 30 to 79, but its percentage operational only went from 28.8 to 53. Maintenance and logistics support improved, but with tight budgets getting in the way, the improvement was not adequate. Costly war readiness spares kits were scarce and several problems were yet to be resolved. A serious flaw in the environmental system ducting pushed the F-111D abort rate above that of other F-111s. Finally, the F-111D's landing gear still needed work, as did several of the Mark II's components. It was improbable that the 27th TFW would be operational before January 1974.

The last F-111D was delivered on February 28, 1973. A total of 96 F-111Ds had been accepted. The Air Force accepted one F-111D in fiscal year (FY) 1970, none the follow-

This four ship of 522nd TFS F-111Ds was photographed over the Great Pyramids of Egypt in August of 1983 (USAF Photo by M. J. Haggerty)

ing fiscal year. Deliveries resumed in July 1971, totaling 28 in FY 1972, and 67 in FY 1973. Sixty RF-111Ds programmed for procurement were canceled in September 1969 in favor of cheaper RF-111As, which were later also canceled.

The F-111Ds spent their full life assigned to the 27th TFW at Cannon AFB. They were retired in 1991 and 1992. The last F-111D retired was 68-0175, which had been last assigned to Sacramento Air Logistics Center. It arrived at AMARC on December 22, 1992. The F-111Ds of the 27th TFW were replaced by F-111Es, F-111Fs and F-111Gs.

The flyaway cost per production aircraft was $8.5 million. This included: airframe – $3,895,000; engines (installed) – $1,229,000; electronics – $2,530,000; ordnance – $6,000; and armament – $844,000. A post-FY 1973 accounting revision showed a decrease of $87,800 in RDT&E for each F-111D. At the same time, it upped the overall price of every F-111D to $13.5 million. This was $188,807 below the unit cost once predicted.

F-111D

The majority of 96 F-111Ds were delivered to the 27th TFW at Cannon AFB, New Mexico, where they spent their entire careers. The F-111D factory numbers were A6-01 through A6-96.

68-0085 (A6-01)
68-0085 first flew on May 15, 1967, and was delivered to the USAF on June 30, 1970. It was used as a test aircraft for both AFFTC at Edwards AFB, and Sacramento Air Logistics Center (SA-ALC) at McClellan AFB. In this photo, taken in September 1971, while assigned to AFFTC, 68-0085 had the white radome with the long flight test pitot boom installed. It was retired on November 13, 1990. It arrived at AMARC after completing 1,359 flights accumulating 2,451.8 flight hours. (Tom Brewer)

68-0086 (A6-02)
68-0086 was delivered to the USAF on July 30, 1971. Like 68-0085, 68-0086 was used as a test aircraft with AFFTC at Edwards, 57th FWW, and the 431st TES at SA-ALC at McClellan until being reassigned to the 27th TFW in July 1991. It is seen here with the WA tail code of the 57th FWW. The aircraft arrived at AMARC on May 6, 1992, with a total of 2,897.1 flight hours. (USAF)

68-0087 (A6-03)
Right: 68-0087 was delivered to the USAF on November 17, 1971. The aircraft was assigned to the 27th TFW at Cannon AFB for its entire career. It's seen here with the CE tail code of the 4427th TFRS. It arrived at AMARC on November 9, 1990, and had flown 1,312 flights accumulating 2,262.1 flight hours. (Marty Isham Collection)

68-0088 (A6-04)
Below: 68-0088 was delivered to the USAF on December 29, 1971. The aircraft was assigned to AFFTC and SA-ALC before being reassigned to the 27th TFW in 1972. It's seen here, photographed on December 29, 1979, with the green tail stripe and the CC tail code of the 481st TFS, 27th TFW. 68-0088 retired from the 27th TFW and arrived at AMARC on June 26, 1992, with a total of 4,393.2 flight hours. (Brian C. Rogers)

68-0089 (A6-05)
68-0089 was delivered to the USAF on August 30, 1971. The aircraft was assigned to AFFTC for its entire career. Its seen here on takeoff at Edwards AFB. 68-0089 arrived at AMARC on November 7, 1990, having flown 1,159 flights and accumulating 2,2118.0 flight hours. (Craig Kaston)

68-0090 (A6-06)
Right: The first F-111D with full Mark II avionics, 68-0090 was delivered to the USAF on October 28, 1971. It arrived at Cannon AFB, and was assigned to the 4427th TFRS of the 27th TFW on November 12, 1971, remaining with the 27th TFW for its entire career. It's seen here, photographed on May 8, 1993, at AMARC with the blue tail stripe and the CC tail code of the 523th FS, 27th FW. The aircraft arrived at AMARC on November 25, 1992. When retired, the aircraft had a total of 4,309.8 flight hours. (Douglas Slowiak/Vortex Photo Graphics)

68-0091 (A6-07)
Below: 68-0091 was delivered to the USAF on December 29, 1971. It is seen here with the green tail stripe and the CC tail code of the 481st TFS, 27th TFW. The aircraft was assigned to the 27th TFW for its entire career and arrived at AMARC on July 10, 1992, with a total of 4,372.2 flight hours. (Terry Love Collection)

68-0092 (A6-08)
68-0092 was delivered to the USAF on January 19, 1972. It is seen here, photographed on October 15, 1980, with the red tail stripe and the CC tail code of the 522nd TFS, 27th TFW. The aircraft was assigned to the 27th TFW for its entire career and arrived at AMARC on May 26, 1992, with a total of 4,398.2 flight hours. (Brian C. Rogers)

68-0093 (A6-09)
Left: The aircraft was delivered to the USAF on January 26, 1972. It's seen here, photographed in April 1973, with a white tail stripe and the CC tail code. 68-0093 belonged to the 524th TFS, 27th TFW at the time the photo was taken. While assigned to the 522nd TFS, 27th TFW, the aircraft crashed and was destroyed on October 3, 1977, on the Melrose bombing range near Cannon AFB, New Mexico. The aircraft, callsign Crazy 46, crashed on the downwind leg of the bombing pattern, setting up for a night radar bomb delivery. The crew, Captains Richard Cardenas and Steven Nelson, did not attempt to eject, and was killed in the crash. The aircraft had a total of 301 flights with 851.5 flight hours when lost. (Don McGarry)

68-0094 (A6-10)
Below: 68-0094 was delivered to the USAF on February 8, 1972. It is seen here, photographed in July 1989, with the yellow tail stripe and the CC tail code of the 524th TFS, 27th TFW. The aircraft was assigned to the 27th TFW for its entire career and arrived at AMARC on September 3, 1992, with a total of 3,958.1 flight hours. (Ben Knowles Collection)

68-0095 (A6-11)
68-0095 was delivered to the USAF on February 29, 1972. It is seen here, photographed in May 1973, in front of the Programmed Depot Maintenance (PDM) hangar at Sacramento Air Materiel Area (later Sacramento Air Logistics Center). It is marked with a white tail stripe and the CC tail code of the 27th TFW. The aircraft experienced a hydraulic fire in the main wheel well fire inflight on February 26, 1976, and was forced to land gear up. The aircraft was shipped to General Dynamics Restoration center at Fort Worth in July 1984, and was repaired on October 12, 1988, and returned to operational status. It remained at Cannon AFB until being retired to AMARC on June 3, 1992. It was retired with a total of 4,388.2 flight hours. (Author)

68-0096 (A6-12)
68-0096 was delivered to the USAF on March 17, 1972. It is seen here, photographed on December 29, 1979, with the yellow tail stripe and the CC tail code of the 524th TFS, 27th TFW. The yellow and black checkerboard below the main stripe indicates it was assigned to Detachment 2 of the 57th FWW. The aircraft was assigned to the 27th TFW for its entire career and arrived at AMARC on September 23, 1992. When retired, the aircraft had a total of 4,702.8 flight hours. (Brian C. Rogers)

68-0097 (A6-13)
68-0097 was delivered to the USAF on April 28, 1972. It is seen here, photographed on July 26, 1984, with the yellow tail stripe and the CC tail code of the 524th TFS, 27th TFW. The aircraft was assigned to the 27th TFW for its entire career and arrived at AMARC on September 11, 1992. When retired, the aircraft had a total of 4,394.2 flight hours. (Charles B. Mayer)

68-0098 (A6-14)
The aircraft was delivered to the USAF on April 26, 1972. It's seen here, photographed on June 16, 1981, with the red tail stripe and the CC tail code of the 522nd TFS, 27th TFW. While assigned to the 27th TFW, the aircraft crashed and was destroyed on June 8, 1988 on the Melrose Range near Cannon AFB. With the TFR not engaged, on the third low level bomb run of the flight, the aircraft hit the ground flat. Captains Glenn Troster and Michael Barritt, were killed.. 68-0098 had logged 1,547 flights and accumulated 3,663.3 flight hours by the time of the crash. (Douglas Slowiak/Vortex Photo Graphics)

CHAPTER THREE: U.S. AIR FORCE TACTICAL F-111s

68-0099 (A6-15)
68-0099 was delivered to the USAF on April 14, 1972. On February 10, 1976, 68-0099 had a 16th stage bleed air duct failure. It was repaired, and the aircraft continued to fly assigned to the 27th TFW until it arrived at AMARC on May 20, 1992. 68-0099 is seen here, photographed in November 1987, with the red tail stripe and the CC tail code of the 522nd TFS, 27th TFW. When retired, the aircraft had a total of 4,394.2 flight hours. (Dave Meehan)

68-0100 (A6-16)
68-0100 was delivered to the USAF on April 28, 1972. The aircraft was assigned to the 27th TFW for its entire career and arrived at AMARC on September 18, 1992. It's seen here at AMARC with the blue tail stripe and the CC tail code of the 523rd TFS, 27th TFW. When retired, the aircraft had a total of 4,393.2 flight hours. (Douglas Slowiak/Vortex Photo Graphics)

68-0101 (A6-17)
68-0101 was delivered to the USAF on March 30, 1972. The aircraft was assigned to the 27th TFW for its entire career. A fire in the liquid oxygen converter severely damaged the aircraft. It was flown by C-5A to General Dynamics, Fort Worth in December 1984, for repair. It was returned to service in December 1986. It's seen here with the red tail stripe and the CC tail code of the 522nd TFS, 27th TFW. 68-0101 arrived at AMARC February 26, 1991. When retired, the aircraft had a total of 3,868.0 flight hours. (Norris Graser)

68-0102 (A6-18)
68-0102 was delivered to the USAF on April 27, 1972. It is seen here, photographed on October 25, 1990, with the yellow tail stripe and the CC tail code of the 524th TFS, 27th TFW. The aircraft was assigned to the 27th TFW for its entire career and arrived at AMARC on May 8, 1992. When retired, the aircraft had a total of 5,004.9 flight hours. (Chris Mayer)

68-0103 (A6-19)
68-0103 was delivered to the USAF on May 16, 1972. It's seen here, photographed in October 1991, marked as a 27th FW Flagship. 68-0103 was assigned to the 27th TFW for its entire career and arrived at AMARC on July 8, 1992. When retired, the aircraft had a total of 4,393.2 flight hours. (Author's Collection)

68-0104 (A6-20)
68-0104 was delivered to the USAF on May 9, 1972. It's seen here, photographed in October 1978, with a CC tail code and a multi-color tail stripe representing all the squadrons of the 27th TFW. The aircraft was assigned to the 27th TFW for its entire career and arrived at AMARC on December 18, 1992. When retired, the aircraft had a total of 4,751.5 flight hours. (Author)

CHAPTER THREE: U.S. AIR FORCE TACTICAL F-111s

68-0105 (A6-21) NO PHOTO AVAILABLE
The aircraft was delivered to the USAF on May 18, 1972. It crashed and was destroyed on March 20, 1973, while assigned to the 27th TFW. The accident occurred near Holbrook, Arizona. The aircraft collided with another F-111D, 68-0158, during a night formation rejoin. In the collision, 0105's wingtip hit 0158's left side near the cockpit. The force of the impact caused the capsule of 0158 to eject; 0105 then ejected. The rocket motor from one capsule burned the main parachute of the other. Neither capsule landed safely, and the crews of both aircraft were killed. The crew of 68-0105 was Majors Richard Brehm and William Halloran. At the time of the crash the aircraft had accumulated 53 flights and 142.7 flight hours at the time of the crash.

68-0106 (A6-22)
Right: 68-0106 was delivered to the USAF on May 25, 1972. It is seen here, photographed on November 12, 1978, with the green tail stripe and the CC tail code of the 481st TFS, 27th TFW. The aircraft was assigned to the 27th TFW for its entire career. It was involved in an accident at Peterson Field, Colorado on August 29, 1980. It arrived at AMARC on September 11, 1992. When retired, the aircraft had a total of 4,390.2 flight hours. (Author's Collection)

68-0107 (A6-23)
68-0107 was delivered to the USAF on May 25, 1972. It is seen here, photographed in May 1985, with the yellow tail stripe and the CC tail code of the 524th TFS, 27th TFW. The aircraft was assigned to the 27th TFW for its entire career and arrived at AMARC on November 19, 1992. When retired, the aircraft had a total of 4,560.5 flight hours. (Terry Love Collection)

68-0108 (A6-24)
68-0108 was delivered to the USAF on May 31, 1972. It is seen here, photographed on May 4, 1991, with the red tail stripe and the CC tail code of the 522nd TFS, 27th TFW. The aircraft was assigned to the 27th TFW for its entire career and arrived at AMARC on September 18, 1992. When retired, the aircraft had total of 4,382.2 flight hours. (Brian C. Rogers)

68-0109 (A6-25)
Right: 68-0109 was delivered to the USAF on June 8, 1972. It is seen here with the yellow tail stripe and the CC tail code of the 524th TFS, 27th TFW. While assigned to the 27th TFW, the aircraft crashed and was destroyed on February 16, 1979. The aircraft lost control during a toss weapons delivery maneuver over Melrose Range, near Cannon AFB, New Mexico. The crew ejected safely. When the aircraft was lost it had accumulated 394 flights and 1,085.5 flight hours. (Terry Love Collection)

68-0110 (A6-26)
Below: The aircraft was delivered to the USAF on June 9, 1972. 68-0110 is seen here, photographed in December 1980, with the red tail stripe and the CC tail code of the 522nd TFS, 27th TFW. It crashed and was destroyed on January 21, 1982, just short of the end of the runway at McClellan AFB, Sacramento, California. The aircraft was on a post Depot Maintenance FCF (Functional Check Flight) when the right engine failed on takeoff. The resulting fire caused a hardover rudder resulting in a loss of aircraft control. The crew, Captains J.V. Leslie and T.S. Sienicki, ejected successfully. 68-0110 had accumulated 574 flights and 1,454.8 flight hours. (Terry Love Collection)

68-0111 (A6-27)
68-0111 was delivered to the USAF on June 23, 1972. It is seen here with the special tail number presentation of 68 F111D, and the yellow tail stripe and the CC tail code of the 524th TFS, 27th TFW. The aircraft was assigned to the 27th TFW for its entire career and arrived at AMARC on February 11, 1991. When retired, the aircraft had a total of 4,324.4 flight hours. (USAF)

68-0112 (A6-28)
68-0112 was delivered to the USAF on June 29, 1972. It is seen here, photographed on October 14, 1980, with the white Giant Voice tail stripe and the CC tail code of the 27th TFW. The aircraft was assigned to the 27th TFW for its entire career and arrived at AMARC on August 5, 1992. When retired, the aircraft had a total of 1381.6 flight hours. (Brian C. Rogers)

68-0113 (A6-29) NO PHOTO AVAILABLE
The aircraft was delivered to the USAF on June 23, 1972. Assigned to the 522nd TFS, 27th TFW, the aircraft crashed and was destroyed on December 21, 1973, on Melrose Range near Cannon AFB, New Mexico. The radar altimeter had been noted as inoperative in preflight, and, with the TFR in standby, the aircraft hit the ground at a shallow angle and skipped. The aircraft hit a second time and the aircrew then attempted ejection. The ejection was not successful. Captain William Delaplane III and Lt Robert Kierce were killed. 68-0113 had accumulated 93 flights and 256.7 flight hours when it crashed.

68-0114 (A6-30)
Right: 68-0114 was delivered to the USAF on July 12, 1972. It is seen here, photographed on June 16, 1981, with the blue tail stripe and the CC tail code of the 523rd TFS, 27th TFW. The aircraft was assigned to the 27th TFW for its entire career and arrived at AMARC on March 23, 1992. When retired, the aircraft had total flight hours of 4,293.8 flight hours. (Brian C. Rogers)

68-0115 (A6-31)
68-0115 was delivered to the USAF on July 17, 1972. It is seen here, photographed on July 29, 1978, with the blue tail stripe and the CC tail code of the 523rd TFS, 27th TFW. The aircraft was assigned to the 27th TFW for its entire career. The aircraft arrived at AMARC on July 15, 1992. When the aircraft was retired, it had a total of 4049.1 flight hours. (Jerry Geer)

68-0116 (A6-32)
68-0116 was delivered to the USAF on July 21, 1972. It is seen here, photographed on January 12, 1980, with the blue tail stripe and the CC tail code of the 523rd TFS, 27th TFW. The aircraft arrived at AMARC on April 24, 1992. When retired, the aircraft had a total of 4388.2 flight hours. (Brian C. Rogers)

68-0117 (A6-33)
68-0117 is seen here, photographed on March 6, 1980, with the CC tail code of the 27th TFW, parked in one of the single aircraft hangars at McClellan AFB. The aircraft was assigned to the 27th TFW for its entire career and arrived at AMARC on February 11, 1991. When retired, the aircraft had a total of 3,953.6 flight hours. (Charles B. Mayer)

68-0118 (A6-34)
Right: 68-0118 was delivered to the USAF on August 4, 1972. It is seen here, photographed on June 16, 1981, with the blue tail stripe and the CC tail code of the 523rd TFS, 27th TFW. The aircraft was assigned to the 431st TES and assigned back to 27th TFW on September 16, 1961. It arrived at AMARC on June 5, 1992. When the aircraft retired, it had a total of 4,391.2 flight hours. (Brian C. Rogers)

68-0119 (A6-35) NO PHOTO AVAILABLE
The aircraft was delivered to the USAF on August 31, 1972. While assigned to the 524th TFTS, 27th TFW at Cannon AFB, the aircraft crashed and was destroyed on February 6, 1980, due to a mid-air collision with Cessna-206A N7393N from Tucumcari, New Mexico. The mid-air occurred at 5800 feet when the Cessna hit the F-111. Ejection occurred at 1300 feet above the ground, but the chute did not deploy in time to prevent the capsule from striking the ground in a nose down attitude. Captain Roy Westerfied and 2 Lt Steven Anderson were killed on impact with the ground. Two civilians in the Cessna were killed. When 68-0119 crashed, it had accumulated 598 flights and 1,513 flight hours.

68-0120 (A6-36)
68-0120 was delivered to the USAF on August 11, 1972. It is seen here, photographed on July 28, 1991, with the blue tail stripe and the CC tail code of the 523rd TFS, 27th TFW. The aircraft was assigned to the 27th TFW for its entire career and arrived at AMARC on September 10, 1991. When retired, the aircraft had a total of 3,415.4 flight hours. (Don Abrahamson)

68-0121 (A6-37)
68-0121 was delivered to the USAF on August 17, 1972. It is seen here, photographed in July 1974, taxiing at McClellan AFB with the red tail stripe and the CC tail code of the 522nd TFS, 27th TFW. The aircraft was assigned to the 27th TFW for its entire career and arrived at AMARC on November 4, 1992. When retired, the aircraft had a total of 4,408.5 flight hours. (Author's Collection)

68-0122 (A6-38)
68-0122 was delivered to the USAF on August 22, 1972. It is seen here, photographed on August 24, 1989, with the red tail stripe and 522nd TFS Flagship markings. The aircraft was assigned to the 27th TFW for its entire career. It arrived at AMARC on November 10, 1992. When retired, the aircraft had a total of 4,388.5 flight hours. (Brian C. Rogers)

68-0123 (A6-39)
68-0123 was delivered to the USAF on September 7, 1972. It is seen here, photographed on November 10, 1976, with the red tail stripe and the CC tail code of the 522nd TFS, 27th TFW. The aircraft arrived at AMARC on November 20, 1992. When retired, the aircraft had a total of 4673.4 flight hours. (Author)
Right: Seen here on May 4, 1986, 68-0123 is marked as the 523rd TFS Flagship. (Pete Wilson)

68-0124 (A6-40)
68-0124 was delivered to the USAF on September 12, 1972. It is seen here, photographed on December 8, 1985, at Carswell AFB with the yellow tail stripe and 524th TFS flagship markings. The aircraft was assigned to the 27th TFW for its entire career and arrived at AMARC on April 25, 1991. When retired, the aircraft had a total of 3,868.4 flight hours. (Brian C. Rogers)

68-0125 (A6-41)
The aircraft was delivered to the USAF on September 18, 1972. 68-0125 is seen here, photographed on June 16, 1991, with the blue tail stripe and the CC tail code of the 523rd TFS, 27th TFW. While assigned to the 27th TFW, the aircraft crashed on September 11, 1987, near Cannon AFB. The aircraft was returning to Cannon AFB with one engine shutdown. The remaining engine decelerated and the shutdown engine could not be restarted in time. The pilot was forced to eject the capsule. The crew ejected safely. When 68-0125 crashed, it had accumulated 1,444 flights and 3,494.2 flight hours. (Brian C. Rogers)

CHAPTER THREE: U.S. AIR FORCE TACTICAL F-111s

68-0126 (A6-42)
Right: 68-0126 was delivered to the USAF on September 26, 1972. It is seen here, photographed on December 29, 1990, with the yellow tail stripe and the CC tail code of the 524th TFS, 27th TFW. The yellow and black checkerboard below the main stripe indicates it was assigned to Detachment 2 of the 57th FWW. The aircraft was assigned to the 27th TFW for its entire career and arrived at AMARC on April 8, 1991. When retired, the aircraft had a total of 3,657.5 flight hours. (Chris Mayer)

68-0127 (A6-43)
68-0127 was delivered to the USAF on September 21, 1972. The aircraft was damaged on July 7, 1981, by a wheel well fire and gear collapse. The aircraft was airlifted to General Dynamics, Forth Worth, using a C-5. The damage to the fuselage center section was repaired using the aft fuselage from F-111F 70-2407. The donor F-111F was the aircraft destroyed by fire before being delivered to the USAF. 68-0127 returned to operational service on September 9, 1983. It is seen here photographed in November 1990 and was marked as the 27th TFW Flagship aircraft. 68-0127 arrived at AMARC on November 25, 1992. When retired, the aircraft had a total of 3,919.1 flight hours. (Ben Knowles)

68-0128 (A6-44)
68-0128 was delivered to the USAF on September 29, 1972. It is seen here, photographed on December 8, 1980, at Nellis AFB with the red tail stripe and the CC tail code of the 522nd TFS, 27th TFW. The aircraft was assigned to the 27th TFW for its entire career and arrived at AMARC on July 1, 1991. When retired, the aircraft had a total of 3,652.4 flight hours. (Terry Love Collection)

68-0129 (A6-45)
68-0129 was delivered to the USAF on October 20, 1972. It is seen here, photographed on January 21, 1980, with the red tail stripe and the CC tail code of the 522nd TFS, 27th TFW. The aircraft was assigned to the 27th TFW for its entire career and arrived at AMARC on March 26, 1991. When retired, the aircraft had a total of 3,620.1 flight hours. (Brian C. Rogers)

68-0130 (A6-46)
The aircraft was delivered to the USAF on October 17, 1972. 68-0130 is seen here, photographed on March 18, 1978, with the red tail stripe and the CC tail code of the 522nd TFS, 27th TFW. It crashed and was destroyed on October 21, 1988, five miles southwest of Cannon AFB, New Mexico. On takeoff from Cannon AFB, the aircrew heard a loud thump. The third stage disk failed causing an inflight fire which burned through the rudder controls causing loss of control. Captains David Swanson and Timothy Gaffney successfully ejected. When 68-0130 crashed, it had accumulated 1,398 flights and 3,322.1 flight hours. (Ben Knowles)

68-0131 (A6-47)
Right: The aircraft was delivered to the USAF on October 27, 1972. 68-0131 is seen here, photographed on February 18, 1984, with the red tail stripe and the CC tail code of the 522nd TFS, 27th TFW. It crashed and was destroyed on August 23, 1990, near Holloman AFB. The aircraft had electrical problems before takeoff, and the left generator control unit (GCU) was replaced; however, the proper post maintenance tests were not accomplished. During flight the electrical problems resurfaced, causing the generator to drop off the line and disrupt the flight controls. The problems also caused uncommanded rolls to the right and eventual loss of control. Majors Richard Davidage, the pilot, and Valdimar Smith, the WSO, ejected safely. When 68-0131 crashed, it had accumulated 1,681 flights and 3,806.2 flight hours. (Brian C. Rogers)

CHAPTER THREE: U.S. AIR FORCE TACTICAL F-111s

68-0132 (A6-48)
Left: The aircraft was delivered to the USAF on November 16, 1972. 68-0132 is seen here, photographed in December 1980, with the red tail stripe and the CC tail code of the 522nd TFS, 27th TFW. While landing at Cannon AFB, the aircraft crashed and was destroyed on March 17, 1988. The aircraft was landing with one engine shut down. While on final approach the good engine was brought to idle, resulting in inadequate hydraulic pressure for the flight controls. The aircraft departed controlled flight, and the air crew successfully ejected. When 68-0132 crashed, it had accumulated 1,376 flights and 3,188.3 flight hours. (Terry Love Collection)

68-0133 (A6-49)
Below: The aircraft was delivered to the USAF on October 25, 1972. As seen here, on October 14, 1980, 68-0133 carried the white Giant Voice Bomb Competition tail stripe of the 27th TFW. While assigned to the 27th TFW, it was heavily damaged due to a birdstrike and the resulting engine fire. The aircraft was written-off and remained at Cannon AFB as an Aircraft Battle Damage Repair (ABDR) training aircraft. The aircraft was scrapped during 1997. (Brian C. Rogers)

68-0134 (A6-50)
68-0134 was delivered to the USAF on October 30, 1972. It is seen here, photographed on December 5, 1989, with the yellow tail stripe and the CC tail code of the 524th TFS, 27th TFW. The aircraft arrived at AMARC on December 16, 1991. When retired, the aircraft had a total of 4,464.4 flight hours. (Brian C. Rogers)

68-0135 (A6-51)
68-0134 was delivered to the USAF on October 26, 1972. It is seen here, photographed on December 5, 1989, with the yellow tail stripe and the CC tail code of the 524th TFS, 27th TFW. The aircraft was assigned to the 27th TFW for its entire career and arrived at AMARC on April 20, 1992. When retired, the aircraft had a total of 4,123.5 flight hours. (Brian C. Rogers)

68-0136 (A6-52)
68-0136 was delivered to the USAF on November 2, 1972. It was assigned to the 27th TFW for its entire career. On September 20, 1979, the second stage of the left engine failed, resulting in an inflight fire. After landing the fire spread, causing major damage. The aircraft was shipped to MASDC, and then on June 21, 1981, to General Dynamics Restoration Facility, Fort Worth. The aft fuselage, center body, and engine nacelles were rebuilt and 68-0136 returned to operational service on May 27, 1983. 68-0136 is seen here, photographed on February 18, 1984, with the red tail stripe and the CC tail code of the 522nd TFS, 27th TFW. It was retired and arrived at AMARC on April 20, 1992. When retired, the aircraft had a total of 3,750.5 flight hours. (Brian C. Rogers)

68-0137 (A6-53)
68-0137 was delivered to the USAF on November 17, 1972. It is seen here, photographed on August 16, 1992, with the blue tail stripe and the CC tail code of the 523rd TFS, 27th TFW. The aircraft was assigned to the 27th TFW for its entire career and arrived at AMARC on November 13, 1992. When retired, the aircraft had a total of 4432.8 flight hours. (Author's Collection)

68-0138 (A6-54)
Right: 68-0138 was delivered to the USAF on November 16, 1972, and was assigned to the 27th TFW for its entire career. 68-0138 is seen here with the red tail stripe and the CC tail code of the 522nd TFS, 27th TFW. It was retired to AMARC in July 1990, arriving after having logged 1,984 flights and 3,993.9 flight hours. (Pat Martin Collection)

68-0139 (A6-55) NO PHOTO AVAILABLE
68-0139 was delivered to the USAF on August 11, 1972. While assigned to 524th TFS, 27th TFW at Cannon AFB, the aircraft, flying as VARK 22, crashed and was destroyed on July 14, 1980. The aircraft was returning with one engine shut down. The crew could not get the aircraft to maintain level flight. The afterburner of the good engine failed and the crew was forced to eject. The module separated successfully from the aircraft, but the main chute failed to open. The module impacted the ground killing the crew, pilot Major Ulysses S. "Sam" Taylor and WSO 1Lt Paul Yeager. The aircraft had logged 495 flights and 1.304.8 flight hours when it crashed.

68-0140 (A6-56)
68-0140 was delivered to the USAF on November 29, 1972. While assigned to the 27th TFW at Cannon AFB, the aircraft was damaged, and written-off in August 1989, due to a birdstrike and the resulting fire. When written off, the aircraft had logged 1,597 flights and 3,772.8 flight hours. 68-0140 is now labeled as "The City of Clovis", and was recently put on a pylon, in Veterans' Park on the west side of Clovis, New Mexico, where U.S. Highway 60/84 enters the city. A memorial located with the aircraft, dedicated November 16, 1996, lists the names of all aircrew members (U.S. and Australian) killed in F-111 losses. (Alec Fushi)

68-0141 (A6-57)
68-0141 was delivered to the USAF on August 24, 1972. It is seen here, photographed on February 18, 1984, with the red tail stripe and the CC tail code of the 522nd TFS, 27th TFW. The aircraft was assigned to the 27th TFW for its entire career and arrived at AMARC on November 5, 1992. When retired, the aircraft had a total of 5,045.2 flight hours. (Doug Remington)

68-0142 (A6-58)
68-0142 was delivered to the USAF on November 30, 1972. It was damaged on landing at Cannon AFB on June 25, 1973, as a result of a damaged (galled) horizontal stabilator actuator. It was repaired and returned to service. 68-0142 is seen here, photographed on July 22, 1990, with the red tail stripe and the CC tail code of the 522nd TFS, 27th TFW. The aircraft was assigned to the 27th TFW for its entire career and arrived at AMARC on November 10, 1992. When retired, the aircraft had a total of 3,894.8 flight hours. (Keith Svendsen)

68-0143 (A6-59)
68-0143 was delivered to the USAF on September 14, 1972. It is seen here, photographed on March 15, 1993, with the blue tail stripe and the CC tail code of the 523th TFS, 27th TFW. The aircraft was assigned to the 27th TFW for its entire career and arrived at AMARC on September 25, 1992. When the aircraft retired, it had a total of 4,459.6 flight hours. (Norris Graser)

68-0144 (A6-60)
68-0144 was delivered to the USAF on December 8, 1972. It is seen here, photographed in April 1981, with the red tail stripe and the CC tail code of the 522nd TFS, 27th TFW. The aircraft was assigned to the 27th TFW for its entire career and arrived at AMARC on April 24, 1992. When retired, the aircraft had a total of 5,442.4 flight hours. (George Cockle)

68-0145 (A6-61)
68-0145 was delivered to the USAF on September 26, 1972. It is seen here, photographed on August 23, 1980, with the red tail stripe and the CC tail code of the 522nd TFS, 27th TFW. The aircraft was assigned to the 27th TFW for its entire career and arrived at AMARC on September 9, 1992. When retired, the aircraft had a total of 5,441.3 flight hours. (Douglas Slowiak/Vortex Photo Graphics)

68-0146 (A6-62) NO PHOTO AVAILABLE
68-0146 was delivered to the USAF on November 30, 1972. It crashed and was destroyed on September 2, 1977, near Cannon AFB. The aircrew had unintentionally left the Flight Control System switch in the Takeoff and Land position. The Flight Manual states that leaving the switch in the Takeoff and Land position rather than the Norm (Normal) position can cause uncommanded roll or pitch oscillations at airspeeds above 400 knots. This caused the aircraft to depart controlled flight during a rudder roll. The crew ejected successfully with both crew members receiving injuries on landing. 68-0146 had logged 259 flights and 743.1 flight hours when it crashed.

68-0147 (A6-63)
Right: 68-0147 was delivered to the USAF on October 17, 1972. It is seen here, photographed on February 18, 1984, with the red tail stripe and marked as the 522nd Flagship. The aircraft was assigned to the 27th TFW for its entire career and arrived at AMARC on July 1, 1991. (Terry Love Collection)

68-0148 (A6-64)
68-0148 was delivered to the USAF on December 21, 1972. The aircraft, badly damaged by a fire resulting from a left engine failure on January 17, 1979, was put in storage at MASDC/AMARC for three years. It was airlifted to General Dynamics Restoration Facility at Fort Worth in mid-1982. The center fuselage and aft fuel tank were repaired, with 68-0148 returning to operational service on December 9, 1983. It is seen here, photographed on February 25, 1984, with the yellow tail stripe and the CC tail code of the 524th TFS, 27th TFW. It was retired and arrived at AMARC on August 28, 1992, with a total accumulation of 3,871.9 flight hours. (Brian C. Rogers)

68-0149 (A6-65)
68-0149 was delivered to the USAF on October 11, 1972. It is seen here, photographed in December 1978, with the red tail stripe and the CC tail code of the 522nd TFS, 27th TFW. The aircraft was assigned to the 27th TFW for its entire career and arrived at AMARC on November 17, 1992. When retired, the aircraft had a total of 3,900.9 flight hours. (Mike Campbell)

68-0150 (A6-66)
68-0150 was delivered to the USAF on December 19, 1972. It is seen here with the blue tail stripe and the CC tail code of the 523th TFS, 27th TFW. The aircraft was assigned to the 27th TFW for its entire career and arrived at AMARC on November 17, 1992. When retired, the aircraft had a total of 3,898.1 flight hours. (Author)

68-0151 (A6-67)
68-0151 was delivered to the USAF on October 25, 1972. It is seen here photographed in October 1991, marked as the 27th TFW Flagship aircraft. The aircraft was assigned to the 27th TFW for its entire career and arrived at AMARC on September 4, 1992. When retired, the aircraft had a total of 3,891.9 flight hours. (Author's Collection)

68-0152 (A6-68)
68-0152 was delivered to the USAF on December 19, 1972. It is seen here with the blue tail stripe and the CC tail code of the 523th TFS, 27th TFW. The aircraft was assigned to the 27th TFW for its entire career and arrived at AMARC on May 13, 1992. When retired, the aircraft had a total of 4,739.9 flight hours. (Terry Love Collection)

68-0153 (A6-69)
68-0153 was delivered to the USAF on November 28, 1972. It is seen here, photographed in September 1990, with the blue tail stripe and the CC tail code marked as the 523th TFS Flagship aircraft. The aircraft was assigned to the 27th TFW for its entire career and arrived at AMARC on November 6, 1992. When retired, the aircraft had total flight hours of 4,455.6. (Jerry Geer)

CHAPTER THREE: U.S. AIR FORCE TACTICAL F-111s

68-0154 (A6-70)
68-0154 was delivered to the USAF on December 29, 1972. It is seen here, photographed in June 1978, with the yellow tail stripe and the CC tail code of the 524th TFS, 27th TFW. The aircraft was assigned to the 27th TFW for its entire career and arrived at AMARC on May 13, 1992. When retired, the aircraft had a total of 4,751.9 flight hours. (Jim Rotramel)

68-0155 (A6-71)
68-0155 was delivered to the USAF on November 30, 1972. It is seen here, photographed on December 6, 1980, with the white Giant Voice Bomb Competition white tail stripe of the 27th TFW. The aircraft was assigned to the 27th TFW for its entire career and arrived at AMARC on June 17, 1992. When retired, the aircraft had a total of 4,613.1 flight hours. (Ray Leader)

68-0156 (A6-72)
68-0156 was delivered to the USAF on December 21, 1972. It is seen here in July 1987, with the blue tail stripe and the CC tail code of the 523th TFS, 27th TFW. The aircraft was assigned to the 27th TFW for its entire career and arrived at AMARC on September 9, 1992. When retired, the aircraft had a total of 4,859.8 flight hours. (Douglas Slowiak/Vortex Photo Graphics)

68-0157 (A6-73)
68-0157 is seen here with the red tail stripe and the CC tail code of the 522nd TFS, 27th TFW. The aircraft was assigned to the 27th TFW for its entire career and arrived at AMARC on June 10, 1992. When retired, the aircraft had a total of 4,684.1 flight hours. (Pat Martin Collection)

68-0158 (A6-74) NO PHOTO AVAILABLE
The aircraft was delivered to the USAF on December 8, 1972. It crashed and destroyed on March 20, 1973 near Holbrook, Arizona following a mid-air collision with F-111D 68-0105 (see 68-0105). Major William Gude and Captain David Blackledge were killed. Almost brand new, 68-0158 had logged 24 flights and 68.1 flight hours when it crashed.

68-0159 (A6-75)
Right: 68-0159 was delivered to the USAF on January 17, 1973. It was assigned to the 27th TFW for its entire career and arrived at AMARC on September 16, 1992. When retired, the aircraft a had total of 4,502.2 flight hours. (Terry Love Collection)

68-0160 (A6-76)
68-0160 was delivered to the USAF on February 14, 1973. It was damaged on landing at Cannon AFB on January 22, 1979. It was shooting an approach with no flaps and no slats following an inflight emergency, when it hit a windshear causing a hard landing. The nose gear collapsed. The aircraft was repaired, and returned to operational service. It crashed and was destroyed on September 14, 1982, near Wagon Mound, New Mexico. The crash occurred at night during an Auto TF descent to a Low Level route. Major Howard Tallman III and Captain William Davis were killed. The aircraft had logged 429 flights and 1,058.8 flight hours when it crashed. (Mike Campbell)

68-0161 (A6-77)
68-0161 was delivered to the USAF on December 22, 1972. It is seen here, photographed on April 25, 1990, with the yellow tail stripe and the CC tail code of the 524th TFS, 27th TFW. The yellow and black checkerboard below the main stripe indicates it was assigned to Detachment 2 of the 57th FWW. The aircraft was assigned to the 27th TFW for its entire career and arrived at AMARC on August 7, 1992. When retired, the aircraft a had total of 3,811.6 flight hours. (Brian C. Rogers)

68-0162 (A6-78)
68-0162 was delivered to the USAF on January 18, 1973. It is seen here, photographed on December 5, 1989, with the yellow tail stripe and the CC tail code of the 524th TFS, 27th TFW. The yellow and black checkerboard below the main stripe indicates it was assigned to Detachment 2 of the 57th FWW. The aircraft was assigned to the 27th TFW for its entire career and arrived at AMARC on July 26, 1991. When retired, the aircraft a had total of 4,264.1 flight hours. (Brian C. Rogers)

68-0163 (A6-79)
68-0163 was delivered to the USAF on January 30, 1973. It is seen here, photographed on October 27, 1990, with the yellow tail stripe and the CC tail code of the 524th TFS, 27th TFW. The aircraft was assigned to the 27th TFW for its entire career and arrived at AMARC on October 2, 1992. When retired, the aircraft had a total of 4,441.2 flight hours. (Chris Mayer)

68-0164 (A6-80)
The aircraft was delivered to the USAF on January 29, 1973. It was named *City of Clovis* and had special marking on both sides of the fuselage behind the cockpit. It crashed and was destroyed on October 17, 1984, near Carrizozo, New Mexico, while assigned to the 27th TFW. The aircraft crashed into a hillside while flying Manual Terrain Following. Auto TFR had been deselected following a fly-up commanded by the automatic TF system The pilot, 1 Lt Albert Torn, and WSO, Captain Alan Payor, were killed. The aircraft had logged 1,272 flights and 3,175,4 flight hours when it crashed. (Author)

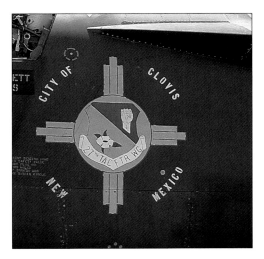

This special City Of Clovis marking was painted on both sides of 68-0164. (Author)

68-0165 (A6-81)
68-0165 was delivered to the USAF on January 30, 1973. It is seen here, photographed in February, 1980, with the yellow tail stripe and the CC tail code of the 524th TFS, 27th TFW. The aircraft was assigned to the 27th TFW for its entire career and arrived at AMARC on November 19, 1992. When retired, the aircraft had a total of 5,741.8 flight hours. (Ben Knowles)

68-0165 had numerous markings applied on its visit to Korea and Australia during October 1979. (Ben Knowles)

68-0166 (A6-82)
68-0166 was delivered to the USAF on February 7, 1973. It is seen here, photographed in February 1980, with the yellow tail stripe and the CC tail code of the 524th TFS, 27th TFW. The aircraft was assigned to the 27th TFW for its entire career and arrived at AMARC on June 12, 1992. When retired, the aircraft had a total of 1,381.6 flight hours. (A.A. Pappas)

CHAPTER THREE: U.S. AIR FORCE TACTICAL F-111s

68-0167 (A6-83) NO PHOTO AVAILABLE
The aircraft was delivered to the USAF on January 30, 1973. It crashed and was destroyed on October 10, 1976, near Artesia, New Mexico. The aircraft had a horizontal stabilator actuator failure causing the aircraft to pitch down. The crew successfully ejected at Mach 1.4 with no injuries, the first successful ejection at that speed. The aircraft had logged 290 flights and 817.6 flight hours when it crashed.

68-0168 (A6-84)
Right: The aircraft was delivered to the USAF on February 6, 1973. 68-0168 is seen here, photographed on June 16, 1981, with the yellow tail stripe and the CC tail code of the 524th TFS, 27th TFW. During flight on September 15, 1990, the aircraft suffered a birdstrike. Birds damaged both engines, the radome, the nose and the cockpit. The #1 engine disintegrated due to damage from the bird strike and the #2 engine was stuck in minimum afterburner. Damage to the cockpit was so severe that a successful ejection may not have been possible, and as a result, the crew elected to attempt to land the aircraft. The aircraft landed successfully with one engine. It was determined not to be cost effective to repair 68-0168, and as a result was written-off at Cannon AFB, New Mexico on March 26, 1990, and was designated a GF-111D battle damage repair trainer. It was later scrapped. 68-0168 had logged 1,850 flights and 4,526.4 flight hours when it was written-off. (Douglas Slowiak/Vortex Photo Graphics)

68-0169 (A6-85)
Below: 68-0169 was delivered to the USAF on February 20, 1973. It is seen here, photographed on October 25, 1990, with the yellow tail stripe and the CC tail code of the 524th TFS, 27th TFW. The aircraft was assigned to the 27th TFW for its entire career and arrived at AMARC on June 19, 1992. When retired, the aircraft had a total of 5,373.6 flight hours. (Chris Mayer)

68-0170 (A6-86)
68-0170 was delivered to the USAF on February 20, 1973. It is seen here, photographed in September 1979, with the yellow tail stripe and the CC tail code of the 524th TFS, 27th TFW. The aircraft was assigned to the 27th TFW for its entire career and arrived at AMARC on May 6, 1992. When retired, the aircraft had a total of 5,004.1 flight hours. (Douglas Slowiak/Vortex Photo Graphics)

GENERAL DYNAMICS F-111 AARDVARK

68-0171 (A6-87)
68-0171 was delivered to the USAF on February 16, 1973. It is seen here, photographed on March 26, 1988, with the yellow tail stripe and the CC tail code of the 524th TFS, 27th TFW. The aircraft was assigned to the 27th TFW for its entire career and arrived at AMARC on September 16, 1991. When retired, the aircraft had a total of 4,235.4 flight hours. (Douglas Slowiak/Vortex Photo Graphics)

68-0172 (A6-88)
68-0163 is seen here with the blue tail stripe and the CC tail code of the 523rd TFS, 27th TFW. The aircraft was assigned to the 27th TFW for its entire career and arrived at AMARC September 23, 1992. When retired, the aircraft had a total flight of 4,280.5 flight hours. (Author's Collection)

68-0173 (A6-89) NO PHOTO AVAILABLE
The aircraft was delivered to the USAF on February 20, 1973. It crashed and was destroyed on November 18, 1978, near Kingman, Arizona. The aircraft departed controlled flight following an explosion and fire in the left engine. The engine failure was due to failure of the ninth stage air seal and the 11th stage compressor disk. The fire burned through the rudder controls causing loss of control. The crew ejected successfully sustaining only minor injuries. The aircraft had logged 658 flights and 1,656.1 flight hours when it crashed.

68-0174 (A6-90)
Right: 68-0174 was delivered to the USAF on August 31, 1972. It is seen here, photographed on May 27, 1991, with the red tail stripe and the CC tail code of the 522nd TFS, 27th TFW. The aircraft was assigned to the 27th TFW for its entire career. On May 21, 1976 the aircraft suffered a fan blade failure during engine run-up prior to takeoff and caught fire. It burned on the runway at Cannon AFB, and then was shipped to MASDC/AMARC. It was shipped from AMARC to General Dynamics Restoration Facility, Fort Worth factory for repair during July 1984. 68-0174 returned to operational service in November 1987. It arrived at AMARC on January 13, 1992. When the aircraft retired, it had a total of 2,217.0 flight hours. (Norris Graser)

68-0175 (A6-91)
68-0175 was delivered to the USAF on September 1, 1972. Seen here, when photographed on August 6, 1975, 68-0175 wears the Sacramento ALC crest on its tail. The aircraft arrived at AMARC on December 28, 1992, after being assigned to Sacramento Air Logistics Center. When retired, the aircraft had a total of 2,456.9 flight hours. (Author)

68-0176 (A6-92)
68-0176 was delivered to the USAF on October 5, 1972. It is seen here on March 3, 1980, with the yellow tail stripe and the CC tail code of the 524th TFS, 27th TFW. It is carrying MK 82s painted orange. The aircraft was assigned to the 27th TFW for its entire career and arrived at AMARC on November 13, 1992. When retired, the aircraft had a total of 4,593.1 flight hours. (Jim Goodall)

68-0177 (A6-93)
68-0177 was delivered to the USAF on October 31, 1972. It was assigned to the 27th TFW for its entire career and arrived at AMARC on November 3, 1992. It is seen here on March 10, 1993, at AMARC in the markings of the 523rd FS. When retired, the aircraft had a total of 4,607.5 flight hours. (Norris Graser)

68-0178 (A6-94)
68-0178 was delivered to the USAF on December 11, 1972. It is seen here with the yellow tail stripe and the CC tail code of the 524th TFS, 27th TFW. In this photo it carries an RAAF insignia on the nose, left over from 68-0178s last visit "down under." The aircraft was assigned to the 27th TFW for its entire career and arrived at AMARC on June 24, 1992. When retired, the aircraft had a total of 1,380.6 flight hours. (Don McGarry)

68-0179 (A6-95)
68-0179 was delivered to the USAF on December 29, 1972. It is seen here, photographed on September 15, 1979, with the yellow tail stripe and the CC tail code of the 524th TFS, 27th TFW. The aircraft was assigned to the 27th TFW for its entire career and arrived at AMARC on September 25, 1992. When retired, the aircraft had a total of 4,910.3 flight hours. (Douglas Slowiak/Vortex Photo Graphics)

68-0180 (A6-96)
68-0180 was delivered to the USAF on January 23, 1973. It is seen here, photographed in January 1980, with the yellow tail stripe and the CC tail code of the 524th TFS, 27th TFW. The aircraft was assigned to the 27th TFW for its entire career and arrived at AMARC on July 17, 1992. When retired, the aircraft had a total of 4,495.2 flight hours. (Terry Love Collection)

F-111E

Although the F-111E was authorized for production after the F-111C, F-111D, and FB-111A, it was the first of these four models to reach an operational capability, beating the FB-111A by one month. The go-ahead decision for the F-111E was issued on February 27, 1968. The decision underscored the F-111's urgency. Since the sophisticated F-111D could not be had quickly, the Air Force had to approve the simpler F-111E configuration for its second tactical wing. Designated F-111E, the aircraft closely resembled the F-111A.

The F-111E was the first production model with Triple Plow II air inlets, improving engine operation at high speed and high altitude; and a stores management set, corresponding to the one planned for the F-111D and F-111F aircraft. Triple Plow II (an improvement of the General Dynamics Triple Plow I air diverter that accompanied the F-111A's P-3 engines) was the main difference between the E and A models. Still, F-111E production was postponed for six months in the beginning. This gave time for F-111A modifications including Harvest Reaper and wing-carry-through box improvements (begun in January 1969) to become part of the F-111E production configuration.

The first delivery of the F-111E occurred on August 20, 1969. The first F-111E was delivered concurrent with delivery of the last F-111As. The aircraft entered operational service on September 30, 1969. TAC's 27th Tactical Fighter Wing at Cannon AFB reached initial operational capability in the fall of 1969. The Wing had 29 F-111Es by December 1969.

Special tests requiring additional equipment on two of the first five F-111Es delayed the program, which had already been affected by production slippages. The Category I and II flight tests started in October 1969, with the first F-111E delivered to Cannon. F-111E Category II system evaluation tests were concluded on July 23, 1971, after showing that the aircraft's major subsystems worked well. Category I separation testing for nuclear weapons was completed in April 1972, with stability and control tests completed in June. The F-111E program slipped another six months, following the December 1969, loss of the 15th F-111A. The Air Force refused to accept any F-111 deliveries until the end of July 1970, when the fleet grounding was lifted. All F-111Es (accepted before and after the grounding) went through the Recovery Program and other structural inspections stemming from the December 1969, accident.

Like the F-111A, the F-111E had an integral radar homing and warning (RHAW) and electronic countermeasures (ECM) capability installed at the factory during production. Aircraft with this capability was greatly needed in Europe. The United States Air Forces Europe (USAFE) counted on the F-111E for the all-weather and night work its F-4s were not equipped to accomplish. Despite the program's initial slippage, the first two of the 79 F-111Es, slated for USAFE's 20th Tactical Fighter Wing, arrived in England on September 11, 1970. These 79 aircraft were out of the total 94 aircraft (counting the five productions allocated to the testing program). The remaining F-111Es initially stayed with TAC. The 442nd squadron at Nellis used them to train F-111 pilots, including USAFE pilots. The 79th TFS, one of the wing's three squadrons, reached Initial Operating Capability (IOC) in December 1970. The 20th Tactical Fighter Wing became fully operational in November 1971.

The F-111E shared most of the operational and support deficiencies of the F-111A. An April 23, 1971, F-111E crash (67-0117), during a Category II flight test, killed its two crew members. In this accident a malfunction of the recovery parachute (part of the escape module) occurred. The recovery parachute problem was immediately fixed in the fleet.

Another F-111E (68-0018) crash landed in Scotland on January 18, 1972. This accident pointed out the need for an audio and visual stall warning system. F-111 pilots could not determine approaching stalls by feel, mistaking rudder pedal's vibrations for airframe buffet. Sacramento Air Materiel Area designed a stall warning change.

The last two of 94 F-111E aircraft were delivered on May 28, 1971. The Air Force accepted 31 F-111Es during fiscal year (FY) 1970 (August through December 1969; none during the following six months). Deliveries resumed in July 1970,

67-0120 is seen here in 57th FWW markings with the wings fully swept to 72 degrees. (USAF)

with 63 F-111Es accepted during FY 1971. The flyaway cost per production aircraft was $9.2 million broken down as follows: airframe – $4,756,000; engines (installed) – $1,511,000; electronics – $1,945,000; ordnance – $7,000; and armament – 1,060,000. Additional costs included $2,826,500 of RDT&E cost and $24,771 worth of modification per aircraft, bringing actual F-111E unit cost to $12,130,271.

By mid-1973, most of the F-111Es assigned to USAFE were combat ready. These aircraft were combat ready prior to the F-111D and FB-111A becoming operational. This out-of-sequence development was not rare. Technical problems often delayed a model's production in favor of a later model in the series. The F-111E continued as part of NATO's arsenal, waging the Cold War from its base in the United Kingdom until 1991.

The older analog avionics of the F-111Es were upgraded as part of the Avionics Modernization Program (AMP). The AMP Es were delivered to the 20th TFW, equipping the 79th TFS first, followed by the 77th TFS, and finally the 55th TFS.

In August 1990, F-111Es were deployed on a training exercise to Incirlik AB, Turkey when Saddam Hussein invaded Kuwait. The detachment remained at Incirlik building up to 22 aircraft by the beginning of the Gulf War. The F-111Es night operations staged from Turkey were known as Operation Proven Force. These operations were conducted from medium altitude without loss. F-111Es provided early support to Provide Comfort from Incirlik following the Gulf War.

With the end of the Cold War, F-111E operations in the United Kingdom were reduced. By mid-July 1993, the AMP modified aircraft had been transferred to the 428th TFTS, 27th Fighter Wing at Cannon AFB replacing the F-111Gs and used as aircrew training aircraft. The E models of the 428th TFTS at Cannon had their engines changed to the updated TF30-P109 also used in the EF-111A. The non-modified AMP F-111Es were retired to AMARC during the second half of 1993. The remainder of the F-111Es, the AMP aircraft, was finally retired to AMARC during the second half of 1995.

F-111E

Most of the 94 F-111Es manufactured spent a majority of their careers assigned to the 20th TFW at RAF Upper Heyford, Oxfordshire, England. The F-111E factory numbers continued from the F-111A numbers and were A1-160 through A1-253. The F-111Es deployed to Incirlik, Turkey for Desert Storm, and returned to Upper Heyford by March 1991. The F-111Es deployed back to the U.S. in mid-1993, with the AMP modified aircraft going to Cannon AFB, New Mexico, and the non-modified aircraft retiring to AMARC.

67-0115 (A1-160) (E-1)
67-0115 was delivered to the USAF on August 30, 1969. It was used by NASA to test the Integrated Propulsion Control System (IPCS) (fly by wire engine controls) between September 4, 1975 and February 27, 1976. While assigned to Eglin AFB, the aircraft was used as a test bed for various systems including the Hughes version of the Pave Mover battlefield surveillance radar in 1983. It was later used as a flight test chase aircraft for the B-1 Strategic Bomber Program. It was retired to AMARC on May 17, 1993, with a total of 2,133.8 flight hours. (USAF)

67-0116 (A1-161) (E-2)
Right: The aircraft was delivered to the USAF on August 30, 1969. 67-0116 seen here on November 2, 1974, was assigned to Eglin AFB. Returning to Eglin AFB on October 27, 1976, the aircraft landed with low fuel, landing too fast and hit too hard. While trying to recover, the pilot induced pitch oscillations after landing which caused loss of control. The pilot then ejected the capsule. The crew escaped without injury. The aircraft crashed and was destroyed. 67-0116 had accumulated 381 flights and 581.8 flight hours. (Tom Brewer Collection)

67-0117 (A1-162) (E-3)
Right: The aircraft was delivered to the USAF on October 13, 1969. This photo of 67-0117 was taken in July 1970, at Edwards AFB. On April 23, 1971, while assigned to AFFTC at Edwards AFB, the aircraft departed controlled flight during a gun firing flight test wind up turn. The aircraft crashed in the Mojave Desert, California. Major James W. Hurt III and Major Robert J. Furman were killed. The parachute failed to deploy because the parachute cover had failed to release. When 67-0117 crashed, it had accumulated 97 flights and 152.7 flight hours. (Bob Trimble)

67-0118 (A1-163) (E-4)
67-0118 was delivered to the USAF on September 19, 1969. It was assigned to Eglin AFB for weapons testing. It's seen here at Luke AFB, Arizona in July 1989. The aircraft suffered damage during June 1981, when a high pressure gas bottle used for inflating the crew module impact attenuation bags exploded. The aircraft was airlifted by C-5A to General Dynamics Restoration Facility, Forth Worth. Extensive rebuilding of the crew module, taking 11 months and costing $1.85 million, was required before the aircraft returned to service in July 1982. The aircraft arrived at AMARC on September 19, 1994. When retired, the aircraft had total of 2,153.8 flight hours. (Bob Shane)

67-118 is seen here on August 28, 1992 wearing the ET tail code of 3246th Test Wing. (Nate Leong)

67-0119 (A1-164) (E-5)
Right: The aircraft was originally assigned to the 442 TFTS, 474th TFW in 1969. It transferred to the 57th FWW for a short time, being reassigned to the 20th TFW at Upper Heyford in 1972. 67-0119 is seen here photographed on June 16, 1992, with the red tail cap of the 77th TFS, 20th TFW. The aircraft left Upper Heyford and arrived at AMARC on August 25,1993. When retired, the aircraft had a total of 5,547.3 flight hours. (Tom Kaminski Collection)

67-0120 (A1-165) (E-6)
67-0120 was delivered to the USAF on October 31, 1969. It was first assigned to Det 3, 57th FWW at Nellis. In 1978 it was reassigned to the 20th TFW at Upper Heyford. The aircraft deployed to Incirlik, Turkey, as part of Operation Proven Force, with mission markings showing 19 Desert Storm scores. It was retired on October 19, 1993, and is presently on display at RAF Duxford Museum. As seen here on June 4, 1988, the aircraft was marked as the 20th TFW commander's aircraft during its active career. It had been nicknamed *The Chief*. (Jerry Geer)

67-0121 (A1-166) (E-7)
67-0121 was delivered to the USAF on November 5, 1969. It was originally assigned to the 442 TFTS, 474th TFW in late 1969. It transferred to the 57th FWW for a short time, as seen here in December 1970, wearing the WF tail code of the 422nd FWS. It was reassigned to the 20th TFW at Upper Heyford in 1972. The aircraft deployed to Incirlik, Turkey, as part of Operation Proven Force, with mission markings showing 21 Desert Storm scores. 67-0121 was nicknamed *Night Stalker*. The aircraft arrived at AMARC on October 21, 1993. When retired, the aircraft had a total of 5,965.7 flight hours. (Tom Brewer Collection)

CHAPTER THREE: U.S. AIR FORCE TACTICAL F-111s

67-0122 (A1-167) (E-8)
67-0122 was delivered to the USAF on November 30, 1969. It was first assigned to the 57th FWW at Nellis AFB, and reassigned to the 20th TFW in 1972. It remained at Upper Heyford until arriving at AMARC on August 25, 1993. When the aircraft retired, it had a total of 5,815.7 flight hours. While at Upper Heyford, it was named *Rowdy Rebel*. (Author's Collection)

67-0123 (A1-168) (E-9)
67-0123 was delivered to the USAF on November 5, 1969. It was first assigned to the 57th FWW at Nellis AFB, and reassigned to the 20th TFW in 1972. 67-0123 is seen here in the summer of 1975, wearing the WA tail code of the 57th FWW. It remained at Upper Heyford until arriving at AMARC on November 4, 1993. When retired, the aircraft had a total of 5,299.8 flight hours. While at Upper Heyford, 67-0123 was nicknamed *The Bold One*. (Author)

67-0124 (A1-169) (E-10)
67-0124 was delivered to the USAF on November 5, 1969. It was first assigned to Det 3, 57th FWW at Nellis. In 1972, 67-0124 was reassigned to ADTC at Eglin AFB. Seen here, photographed in August 1972, 67-0124, on the runway with F-111D 68-0033 wingman, carries the WA tail code of the 57th FWW. It was later assigned to the 431st TES, retiring to AMARC on May 27, 1992. When the aircraft retired, it had a total of 3,446.1 flight hours. (Mick Roth)

68-0001 (A1-170) (E-11)
68-0001 was delivered to the USAF on November 26, 1969. Seen here in a photo taken in September 1971, 68-0001 carries the 77th TFS squadron tail code JT. While assigned to the 20th TFW at Upper Heyford, the aircraft crashed and was destroyed on February 5, 1990 on Wainfleet Range, The Wash, north of Milton Keynes, England. While in a steep descending turn, without TF engaged, the wingtip caught the water. The aircraft cartwheeled into the water, killing the crew, Captain Clifford Massengill and 1Lt Thomas Dorsett. When 68-0001 crashed, it had accumulated 1,641 flights and 4,321.3 flight hours. (Pat Martin Collection)

68-0002 (A1-171) (E-12)
68-0002 was delivered to the USAF on December 2, 1969. It was originally assigned to the 442 TFTS, 474th TFW in late 1969, as seen in this photo taken on February 19, 1972. It was reassigned to the 20th TFW at Upper Heyford in 1972. During 1987 and 1988 while assigned to the 20th TFW 68-0002 was nicknamed *Imperial Wizard*. The aircraft arrived at AMARC on December 16, 1992. When retired, the aircraft had a total of 5,701.1 flight hours. (Douglas Slowiak Collection)

68-0003 (A1-172) (E-13)
68-0003 was delivered to the USAF on November 25, 1969, and was initially assigned to the 442 TFTS, 474th TFW in late 1969. It transferred to the 20th TFW at Upper Heyford in 1972. While assigned to the 20th TFW at Upper Heyford, the aircraft crashed and was destroyed on December 19, 1979 at Craignaw, Scotland. The aircraft hit a low hill while in a low level tactical turn. Captain Richard A. Hetzner and Captain Raymond C. Spaulding, did not attempt ejection and were killed. 68-0003 had accumulated 800 flights and 2,072.6 flight hours when it crashed. (Terry Love Collection)

CHAPTER THREE: U.S. AIR FORCE TACTICAL F-111s

68-0004 (A1-173) (E-14)
68-0004 was delivered to the USAF on November 6, 1969. It deployed to Incirlik, Turkey as part of Operation Proven Force and marked with 6 Desert Storm scores. The aircraft arrived at AMARC on October 28, 1993. When the aircraft retired, it had a total of 6,011.1 flight hours. (Douglas Slowiak Collection)

68-0005 (A1-174) (E-15)
68-0005 was delivered to the USAF on December 2, 1969. It is seen here, at Nellis AFB in February 1983, wearing the black and white checker board tail cap of the 55th TFS, 20th TFW. The aircraft deployed to Incirlik, Turkey, as part of Operation Proven Force and was marked with 29 scores. 68-0005 carried the nickname *Born in the USA*. The aircraft arrived at AMARC on October 21, 1993. When retired, the aircraft had a total of 5,719.9 flight hours. (Kevin Patrick)

68-0006 (A1-175) (E-16)
68-0006 was delivered to the USAF on November 25, 1969. It is seen here in August 1976, wearing the yellow tail cap of the 79th TFS, 20th TFW. While operating with the 20th TFW, 68-0006 was named *Free Bird*. The aircraft arrived at AMARC on November 3, 1993. When retired, the aircraft had a total of 5,854.1 flight hours. (Craig Kaston Collection)

68-0007 (A1-176) (E-17)
68-0007 was delivered to the USAF on November 30, 1969. It arrived at AMARC on October 28, 1993. When retired, the aircraft had a total of 5,686.1 flight hours. (Douglas Slowiak Collection)

68-0008 (A1-177) (E-18) NO PHOTO AVAILABLE
The aircraft was delivered to the USAF on December 18, 1969. While assigned to the 20th TFW, 68-0008 crashed and was destroyed on May 15, 1973. The aircraft hit a bird near Macrihanish in western Scotland. The birdstrike caused an engine fire and loss of control. The crew, pilot Captain Andy Peloquin and WSO Captain Alvarez, ejected successfully. When the aircraft crashed, it had accumulated 138 flights and 4,13.6 flight hours.

68-0009 (A1-178) (E-19)
68-0009 was delivered to the USAF on November 26, 1969. It is seen here on August 14, 1982, wearing the red tail cap of the 77th TFS, 20th TFW. It arrived at AMARC on May 11, 1992, and was the first F-111E retired. When 68-0009 retired, it had a total of 5,430.8 flight hours. (Author's Collection)

68-0010 (A1-179) (E-20)
68-0010 was delivered to the USAF on December 2, 1969. It is seen here on March 6, 1989, wearing the yellow and black tail cap of the 79th TFS Tigers. The aircraft arrived at AMARC on November 16, 1992. When 68-0010 retired, it had a total of 5,654.8 flight hours. (Tom Kaminski Collection)

CHAPTER THREE: U.S. AIR FORCE TACTICAL F-111s

68-0011 (A1-180) (E-21)
68-0011 was delivered to the USAF on December 8, 1969. It is seen here in May, 1979, with the yellow tail cap of the 79th TFS, 20th TFW. After being retired on June 24, 1962, it was placed on display as a gate guard at RAF Lakenheath Museum marked as the 48th TFW commanders aircraft. (Douglas Slowiak Collection)

68-0012 (A1-181) (E-22) NO PHOTO AVAILABLE
68-0012 was delivered to the USAF on December 13, 1969. While assigned to the 20th TFW, it crashed and was destroyed on October 30, 1979, at RAF Alconbury, UK. While in the holding pattern for landing, the airspeed was allowed to decay. When the airspeed became too low, the aircraft departed controlled flight. The crew ejected successfully. The aircraft had accumulated 711 flights and 2,017.2 flight hours.

68-0013 (A1-182) (E-23)
68-0013 was delivered to the USAF on November 26, 1969. It deployed to Incirlik, Turkey as part of Operation Proven Force and carried 24 scores. While assigned to Upper Heyford 68-0013, was named *Excalibur*. The aircraft arrived at AMARC on November 12, 1992. It's seen here at AMARC in March 1993. When retired, the aircraft had a total of 5,522.7 flight hours. (Douglas Slowiak/Vortex Photo Graphics)

68-0014 (A1-183) (E-24)
68-0014 was delivered to the USAF on December 18, 1969. It is seen here, photographed on June 23, 1988, wearing the blue tail cap of the 55th TFS. It arrived at AMARC on June 26, 1992. When retired, the aircraft had a total of 5,931.4 flight hours. (Rolf Flinzner)

68-0015 (A1-184) (E-25)
68-0015 was delivered to the USAF on November 26, 1969. It is seen here at Nellis AFB on January 16, 1976, on its way to Programmed Depot Maintenance at McClellan AFB. The aircraft deployed to Incirlik, Turkey as part of Operation Proven Force. 68-0015 was named *Ozone Ranger* while assigned to Upper Heyford. The aircraft arrived at AMARC on April 2, 1993. When retired, the aircraft had a total of 5,425.0 flight hours. (Author)

68-0016 (A1-185) (E-26)
68-0016 was delivered to the USAF on November 30, 1969. It is seen here in November 1972, wearing the JS squadron tail code on the 55th TFS. The aircraft deployed to Incirlik, Turkey as part of Operation Proven Force. 68-0016 arrived at AMARC on April 2, 1993. When retired, the aircraft had a total of 5,569.7 flight hours. (Gerry Markgraf Collection)

68-0017 (A1-186) (E-27)
68-0017 was delivered to the USAF on November 30, 1969. It deployed to Incirlik, Turkey as part of Operation Proven Force. The aircraft arrived at AMARC on December 16, 1992. When retired, the aircraft had a total of 5,505.2 flight hours. (Tom Brewer Collection)

68-0018 (A1-187) (E-28) NO PHOTO AVAILABLE
68-0018 was delivered to the USAF on December 13, 1969. While assigned to the 20th TFW, it crashed and was destroyed on January 18, 1972, while on landing approach to RAF Leuchars, Scotland. The aircraft was on base leg of a ground controlled radar approach (GCA) with the wings set at 35 degrees. As the aircraft slowed, it stalled and crashed near Coupar Angus, Scotland. Lt Col Floyd B. Sweet and Lt Col Kenneth T. Blank did not attempt ejection and were killed. When the aircraft crashed, it had flown 141 flights accumulating 362.9 flight hours.

CHAPTER THREE: U.S. AIR FORCE TACTICAL F-111s

68-0019 (A1-188) (E-29)
The aircraft was delivered to the USAF on November 30, 1969. It's seen here at Upper Heyford on January 7, 1977. While assigned to the 20th TFW, 68-0019, flying near Kinloss, United Kingdom, crashed and was destroyed on August 9, 1984, on the edge of Loch Eye near Fearn, Scotland. The aircraft hit a bird which destroyed the radome and damaged an engine. In addition, the birdstrike caused the Angle of Attack and Stall Inhibitor System to fail. As a result, the aircraft departed controlled flight. The crew, Pilot Captain Ralph Jodice and WSO Captain Paul Emrich, ejected safely. The aircraft had accumulated 1,160 flights and 3,076.6 flight hours. (Michael France)

68-0020 (A1-189) (E-30)
68-0020 was delivered to the USAF on December 23, 1969. Seen here in May of 1990, it is marked as a 20th TFW Flagship aircraft. It was retired in December 1993, and is now on display at the Hill AFB Museum, Utah. (Author's Collection)

68-0021 (A1-190) (E-31)
68-0021 was delivered to the USAF on December 17, 1969. It is seen here on October 1983, wearing the yellow tail cap of the 79th TFS, 20th TFW. The aircraft arrived at AMARC on July 6, 1993. When 86-0021 retired, it had a total of 5,421.0 flight hours. (Scott Wilson)

68-0022 (A1-191) (E-32)
68-0022 was delivered to the USAF on September 30, 1970. It is seen here on August 7, 1976, wearing the blue tail cap of the 55th TFS, 20th TFW. After spending most of its career at Upper Heyford, where it was nicknamed *Thundercat*, 68-0022 transferred to the 428th FS, 27th FW at Cannon, AFB, New Mexico in December 1993. It was also used at Cannon as a ground trainer. The aircraft arrived at AMARC on December 15, 1995. When retired, the aircraft, had a total of 5,768.0 flight hours. (Author's Collection)

68-0023 (A1-192) (E-33)
68-0023 was delivered to the USAF on December 23, 1969. It is seen here on May 4, 1979, wearing the yellow tail cap of the 77th TFS, 20th TFW. While assigned to the 20th TFW, 68-0023 was nicknamed *Aces High*. The aircraft arrived at AMARC on October 13, 1993. When retired, the aircraft had a total of 5,605.6 flight hours. (John Owen)

68-0024 (A1-193) (E-34) NO PHOTO AVAILABLE
68-0024 was delivered to the USAF on October 25, 1970. While assigned to the 20th TFW, 68-0024 crashed and was destroyed on January 11, 1973, near RAF Upper Heyford. A broken B-nut on the fuel manifold caused a fire which eventually burned through the rudder control rod. The crew attempted to recover the aircraft, but when the landing gear was lowered, aircraft control was lost. The pilot, Major Robert Kroos, and WSO, Major Roger Beck, ejected successfully. The aircraft had accumulated 34 flights and 522.2 flight hours when it crashed.

68-0025 (A1-194) (E-35)
Right: 68-0025 was delivered to the USAF on September 30, 1970. It is seen here in June 1988, wearing the blue tail cap of the 55th TFS, 20th TFW. Assigned to the 20th TFW for its entire career, 68-0025 arrived at AMARC on October 13, 1993. When retired, the aircraft had a total of 5,744.7 flight hours. (Kevin Foy)

CHAPTER THREE: U.S. AIR FORCE TACTICAL F-111s

68-0026 (A1-195) (E-36)
68-0026 was delivered to the USAF on May 28, 1971. It is seen here in July 1993, wearing the blue tail cap of the 55th TFS, 20th TFW. Twenty-three Desert Storm mission scores are visible below the windshield. The aircraft deployed to Incirlik, Turkey as part of Operation Proven Force. 68-0026 arrived at AMARC on October 21, 1993. When retired, the aircraft had a total of 5,868.2 flight hours. (Tom Kaminski Collection)

68-0027 (A1-196) (E-37)
68-0027 was delivered to the USAF on December 20, 1970. It is seen wearing the red tail cap of the 77th TFS, 20th TFW. After assignment to the 20th TFW at Upper Heyford, AMP modified 68-0027 was assigned to Sheppard AFB, Texas as a ground training airframe (GF-111E). (Pat Martin Collection)

68-0028 (A1-197) (E-38)
68-0028 was delivered to the USAF on October 28, 1970. As seen here in this photograph taken in August 1976, 68-0028 was painted US Bicentennial markings named "Spirit of 76". The aircraft arrived at AMARC on September 2, 1993. When retired, the aircraft had a total of 5,416.3 flight hours. (Douglas Slowiak Collection)

68-0029 (A1-198) (E-39)
68-0029 was delivered to the USAF on October 31, 1970. It is seen here in August 1991, wearing the black tail cap of the 55th TFS, 20th TFW. The aircraft deployed to Incirlik, Turkey as part of Operation Proven Force. 68-0029 flew missions resulting in it being marked with 18 scores. The aircraft arrived at AMARC on November 18, 1993. When retired, the aircraft had a total of 5,771.4 flight hours. (Ben Knowles Collection)

68-0030 (A1-199) (E-40)
68-0030 was delivered to the USAF on October 26, 1970. It is seen here in June 1993, wearing the blue tail cap of the 55th TFS, 20th TFW. Nicknamed *Top Cat*, 68-0030 spent its career assigned to the 20th TFW at Upper Heyford. The aircraft arrived at AMARC on November 3, 1993. When retired, the aircraft had a total of 5,631.1 flight hours. (Tom Kaminski Collection)

68-0031 (A1-200) (E-41)
68-0031 was delivered to the USAF on October 28, 1970. It is seen here on April 18, 1984, wearing the yellow tail cap of the 79th TFS, 20th TFW. The aircraft deployed to Incirlik, Turkey as part of Operation Proven Force. 68-0031 arrived at AMARC on August 12, 1993. When retired, the aircraft had a total of 5,498.6 flight hours. (Scott Wilson)

CHAPTER THREE: U.S. AIR FORCE TACTICAL F-111s

68-0032 (A1-201) (E-42)
68-0032 was delivered to the USAF on October 31, 1970. Nicknamed *Kitty*, 68-0032 spent the majority of its career assigned to the 20th TFW at Upper Heyford. 68-0032 was an AMP modified aircraft and was assigned to the 428th FS at Cannon in April 1993. It is seen here at Cannon AFB in May 1993, wearing the blue tail stripe of the 428th FS Buccaneers of the 27th FW. The aircraft arrived at AMARC on October 11, 1995. When retired, the aircraft had a total of 6,263.6 flight hours. (Alec Fushi)

68-0033 (A1-202) (E-43)
Right: 68-0033 was delivered to the USAF on August 21, 1970. It After initially assigned to 422nd FWS, as seen here on June 13, 1975, with WA tail codes of the 57th FWW. 68-0033 was transferred to the 20th TFW in 1978. Nicknamed *Hat Trick*, it spent the remainder of its career assigned to the 20th TFW at Upper Heyford. The aircraft was retired on April 22, 1993, and is on display at Pima Air and Space Museum (PASM) Tucson, Arizona. 68-0033 is painted with the Upper Heyford markings with 28 mission score marks painted on the fuselage. The aircraft arrived at AMARC on December 13, 1995. When retired, it had a total of 4,687.0 flight hours. (Bill Schell)

68-0034 (A1-203) (E-44)
Below: 68-0034 was delivered to the USAF on August 16, 1970. It is seen here on May 21, 1992, wearing the red tail cap of the 77th TFS, 20th TFW. Nicknamed *Drunken Buzzard*, 68-0032 spent its career assigned to the 20th TFW at Upper Heyford. The aircraft arrived at AMARC on November 17, 1993. When retired, the aircraft had a total of 5,562.2 flight hours. (Terry Love Collection)

68-0035 (A1-204) (E-45)
68-0035 was delivered to the USAF on August 16, 1970. It is seen here on August 14, 1982, wearing the yellow tail cap of the 79th TFS, 20th TFW. Nicknamed *Shamrock Kid*, 68-0035 spent its career assigned to the 20th TFW at Upper Heyford. The aircraft arrived at AMARC on April 26, 1993. When retired, the aircraft had a total of 5,173.0 flight hours. (G.J. Booma)

68-0036 (A1-205) (E-46)
68-0036 was delivered to the USAF on August 10, 1970. It is seen here in July 1987, wearing the black and white checked tail cap of the 55th TFS, 20th TFW. Nicknamed *Wild Fire*, 68-0036 spent its career assigned to the 20th TFW at Upper Heyford. It flew missions in support of Desert Storm and recorded 32 scores. The aircraft arrived at AMARC on December 16, 1992. When retired, the aircraft had a total of 5,832.9 flight hours. (Author's Collection)

68-0037 (A1-206) (E-47)
68-0037 was delivered to the USAF on June 31, 1970. It is seen here on July 1985, wearing the red tail cap of the 77th TFS, 20th TFW. The aircraft arrived at AMARC on November 4, 1993. When retired, the aircraft had a total of 5,841.8 flight hours. (Terry Love Collection)

CHAPTER THREE: U.S. AIR FORCE TACTICAL F-111s

68-0038 (A1-207) (E-48)
68-0038 was delivered to the USAF on August 13, 1970. It is seen here on May 21, 1992, wearing the yellow tail cap of the 79th TFS, 20th TFW. The aircraft arrived at AMARC on July 6, 1993. When retired, the aircraft had a total of 5255.1 flight hours. (Tom Kaminski Collection)

68-0039 (A1-208) (E-49)
68-0039 was delivered to the USAF on August 9, 1970. It is seen here on April 21, 1994, wearing the red tail cap of the 77th TFS, 20th TFW. Nicknamed *The Baghdad Express*, 68-0039 spent its career assigned to the 20th TFW at Upper Heyford. The aircraft deployed to Incirlik, Turkey as part of Operation Proven Force. It flew missions in support of Desert Storm and recorded 21 scores. The aircraft arrived at AMARC on August 25, 1992. When retired, the aircraft had a total of 5697.8 flight hours. (Scott Wilson)

68-0040 (A1-209) (E-50)
The aircraft was delivered to the USAF on August 10, 1970. 68-0040 was nicknamed *The Other Woman (1987-88)* at Upper Heyford and *Magnificent Marsha (1988 -89)* while assigned to Sacramento ALC. 68-0040 is seen here wearing the yellow and black tiger striped tail cap of the 79th TFS, 20th TFW. The aircraft deployed to Incirlik, Turkey as part of Operation Proven Force. As an AMP modified aircraft, it was reassigned to the 428th FS, 27th FW. While assigned to the 27th TFW at Cannon AFB, New Mexico, 68-0040 became the last F-111E loss when it crashed on February 16, 1995. It was on final approach to Cannon AFB, practicing a single engine approach with retracted flaps, when the "good" engine failed. The pilot immediately applied power to the simulated lost engine, but the engine was not able to spool up in time to prevent the aircraft from contacting the ground. The crew ejected, with the capsule hitting the ground after one swing from the main parachute. The aircraft burned on impact. The crew, pilot Ed "Fig" Newton and WSO 1LT Randolph Winge, received only minor injuries. (Author's Collection)

68-0041 (A1-210) (E-51)
68-0041 was delivered to the USAF on August 10, 1970. It is seen here in November 1982, wearing the yellow tail cap of the 79th TFS, 20th TFW. 68-0041 spent most of its career assigned to the 20th TFW at Upper Heyford. With the closing of Upper Heyford, 68-0041, an AMP modified aircraft was reassigned to the 428th FS, 27th FW. The aircraft arrived at AMARC on September 22, 1994. When retired, the aircraft had a total of 5,524.1 flight hours. (Author's Collection)

68-0042 (A1-211) (E-52)
The aircraft was delivered to the USAF on October 28, 1970. 68-0042 is seen here wearing the JR squadron tail code of the 79th TFS, 20th TFW. While assigned to the 20th TFW, at Upper Heyford, 68-0042 crashed and was destroyed on July 24, 1979 near Leconfield, England. Below weather minimums, the aircraft hit the water in the final turn. The air crew Captain David W. Powell and Captain Douglas A. Pearce, did not try to eject and were killed. The aircraft had accumulated 714 flights and 1,928.6 flight hours. (Douglas Slowiak Collection)

68-0043 (A1-212) (E-53)
68-0043 was delivered to the USAF on August 19, 1970. It is seen here on May 4, 1979, wearing the yellow tail cap of the 79th TFS, 20th TFW. The aircraft was spent most of its career assigned to the 20th TFW at Upper Heyford. 68-0043 arrived at AMARC on October 13, 1993. When retired, the aircraft had a total of 6,110.6 flight hours. (John Owen)

CHAPTER THREE: U.S. AIR FORCE TACTICAL F-111s

68-0044 (A1-213) (E-54)
68-0044 was delivered to the USAF on September 30, 1970. It is seen here on November 25, 1970, wearing the UR squadron tail code of the 79th TFS, 20th TFW. The aircraft spent most of its career assigned to the 20th TFW at Upper Heyford. With the closing of Upper Heyford, AMP modified 68-044 was reassigned to the 428th FS, 27th FW on May 10, 1993. The aircraft arrived at AMARC on October 11, 1995. When retired, the aircraft had a total of 5,649.1 flight hours. (Author's Collection)

68-0045 (A1-214) (E-55)
68-0045 was delivered to the USAF on August 20, 1970. It is seen here in June 1979, wearing the red tail cap of the 77th TFS, 20th TFW. While assigned to the 20th TFW, 68-0045 crashed and was destroyed on December 12, 1979 near Boston (Linc.) UK. The aircraft was lost after range entry, when it plowed deep into coastal mud. No ejection was attempted. Captain Randolph Gaspard and Major Frank Slusher were killed. When the aircraft crashed, it had accumulated 766 flights and 2,126.9 flight hours. (Douglas Slowiak Collection)

68-0046 (A1-215) (E-56)
Right: 68-0046 was delivered to the USAF on August 28, 1970. It is seen here in June 1992, wearing the red tail cap of the 77th TFS, 20th TFW. The aircraft deployed to Incirlik, Turkey as part of Operation Proven Force. 68-0046 flew missions in support of Desert Storm and recorded 12 scores. The aircraft arrived at AMARC on August 12, 1993. When retired, the aircraft had a total of 5,290.7 flight hours. (Tom Kaminski Collection)

68-0047 (A1-216) (E-57)
68-0047 was delivered to the USAF on August 31, 1970. It is seen here in June 1972, wearing the JR squadron tail code of the 79th TFS, 20th TFW. Nicknamed *Til We Meet Again*, 68-0047 spent its career assigned to the 20th TFW at Upper Heyford. The aircraft, an AMP modified F-111E, was reassigned to the 428th FS, 27th TFW on May 10, 1993. The aircraft arrived at AMARC on December 13, 1995, with a total of 5,738.9 flight hours. (Terry Love Collection)

68-0048 (A1-217) (E-58)
68-0048 was delivered to the USAF on September 20, 1970. The aircraft spent most of its career assigned to the 20th TFW at Upper Heyford. As an AMP modified F-111E, 68-0048 was reassigned to the 428th FS, 27th FW. As seen in this photograph taken at Langley AFB on February 26, 1994, 68-0048 was marked as the 428th FS Flagship aircraft. The aircraft arrived at AMARC on October 11, 1995. When the aircraft retired, it had a total of 5,354.3 flight hours. (Brian C. Rogers)

68-0049 (A1-218) (E-59)
68-0049 was delivered to the USAF on January 14, 1971. Nicknamed *Grim Reaper*, 68-0039 spent its career assigned to the 20th TFW at Upper Heyford. The aircraft deployed to Incirlik, Turkey as part of Operation Proven Force. It flew missions in support of Desert Storm and recorded 12 scores. Seen here in September 1991, 68-0049 is marked as the 79th TFS/AMU Flagship aircraft. (Tom Kaminski Collection)

Above and three below: 68-0049 aircraft arrived at AMARC on October 13, 1995, and carried special markings in red and white and was nicknamed *77th Gamblers Last Deal*. When retired, the aircraft had a total of 5454.4 flight hours. (Doug Slowiak/Vortex Photo Graphics)

68-0050 (A1-219) (E-60)
68-0050 was delivered to the USAF on September 22, 1970. It suffered a LOX (liquid oxygen) bottle explosion while on the ramp. The aircraft was rebuilt at General Dynamics Restoration Facility, Fort Worth starting in September 1986, and returned to service at Upper Heyford in November 1988. 68-0050 deployed to Incirlik, Turkey, as part of Operation Proven Force. It is reported to be the only AMP aircraft taking part in Desert Storm. It was reassigned to the 428th FS, 27th FW on April 22, 1993. 68-0050 is seen here in April 1994, wearing the blue tail stripe of the 428th FS, 27th FW. It arrived at AMARC on October 11, 1995. When retired, the aircraft had a total of 4,815.3 flight hours. (Author)

68-0051 (A1-220) (E-61)
68-0051 was delivered to the USAF on September 29, 1970. It is seen here on May 21, 1992, wearing the yellow tail cap of the 79th TFS, 20th TFW. The aircraft spent most of its career assigned to the 20th TFW at Upper Heyford. 68-0051 arrived at AMARC on August 12, 1993. When retired, the aircraft had a total of 5,556.6 flight hours. (Tom Kaminski Collection)

68-0052 (A1-221) (E-62)
The aircraft was delivered to the USAF on September 30, 1970. 68-0052 is seen here in March 1988, wearing the yellow and black tiger stripe tail cap of the 79th TFS, 20th TFW. It is carrying nose art and the nickname *Mr. Gilard*. Also nicknamed *On Guard*, 68-0052 spent its career assigned to the 20th TFW at Upper Heyford. On September 17, 1992, while flying S turns, inbound for a no flap/no slat landing approach, the aircraft lost speed and altitude and impacted the approach lights. The ejection was attempted after hitting the ground. The capsule was damaged in the crash and failed to gain enough altitude for full chute deployment. The pilot, Captain Jerry Lindh, and WSO, Major Michael McGuire, were killed. (Kevin Patrick)

68-0053 (A1-222) (E-63)
68-0053 was delivered to the USAF on October 29, 1970. It is seen here in August 1976, wearing the blue tail cap of the 55th TFS, 20th TFW. The aircraft spent most of its career assigned to the 20th TFW at Upper Heyford. 68-0053 arrived at AMARC on June 26, 1992. When retired, the aircraft had a total of 5,681.0 flight hours. (Douglas Slowiak Collection)

CHAPTER THREE: U.S. AIR FORCE TACTICAL F-111s

68-0054 (A1-223) (E-64)
68-0054 was delivered to the USAF on September 29, 1970. It is seen here wearing the red tail cap of the 77th TFS, 20th TFW. The aircraft spent most of its career assigned to the 20th TFW at Upper Heyford. 68-0054, an AMP modified F-111E, was reassigned to the 428th FS, 27th FW, and arrived at AMARC on December 19, 1995. When retired, the aircraft had a total of 5,585.7 flight hours. (Craig Kaston Collection)

68-0055 (A1-224) (E-65)
68-0055 was delivered to the USAF on October 30, 1970. It spent most of its career assigned to the 20th TFW at Upper Heyford. 68-0055 is seen here on October 24, 1994, wearing the black and white checkerboard tail cap of the 55th TFS, and the nose art *Heartbreaker*. It is marked as the 55th TFS/FS Flagship aircraft. 68-0055 was retired on December 7, 1993, and is now on display at Robins AFB, Georgia. (Ray Leader)

68-0056 (A1-225) (E-66)
68-0056 was delivered to the USAF on October 22, 1970. It is seen here with the black tail cap of the 55th TFS, 20th TFW. The aircraft spent most of its career assigned to the 20th TFW at Upper Heyford. 68-0056 arrived at AMARC on October 21, 1993. When retired, the aircraft had a total of 5,273.7 flight hours. (Ben Knowles Collection)

68-0057 (A1-226) (E-67)
The aircraft was delivered to the USAF on November 21, 1970. 68-0045 is seen here in September 1972, wearing JT, the 77th TFS squadron tail code. While assigned to the 20th TFW, 68-0057 crashed and was destroyed on April 29, 1980, near RAF Bentwaters. The spoilers extended during a formation descent in the weather, and the aircrew lost control and ejected. The pilot, Captain Jack A. Hines and WSO, Captain Richard J. Franks, were killed. When lost, the aircraft had accumulated 801 flights and 2,241.2 flight hours. (Author's Collection)

68-0058 (A1-227) (E-68)
68-0058 was delivered to the USAF on May 14, 1971. Initially assigned to the 422nd FWS, it was reassigned to the 3246th Test Wing at Eglin AFB and used as a test bed for various systems including the Hughes Aircraft Pave Mover battlefield surveillance radar. It is seen here at Eglin AFB on August 17, 1992, with the ET tail code of the 3246th Test Wing. After being retired November 8, 1992, 68-0058 was put on display at USAF Armament Museum, Eglin AFB, Florida. (Brian C. Rogers)

68-0059 (A1-228) (E-69)
68-0059 was delivered to the USAF on November 30, 1970. It is seen here at Nellis AFB in February 1988, wearing the red tail cap of the 77th TFS, 20th TFW. The art named *The Mad Bomber* is visible on the nose. The aircraft spent most of its career assigned to the 20th TFW at Upper Heyford. 68-0059 arrived at AMARC on October 28, 1993. When retired, the aircraft had a total of 5,442.7 flight hours. (Kevin Patrick)

CHAPTER THREE: U.S. AIR FORCE TACTICAL F-111s

68-0060 (A1-229) (E-70)
The aircraft was delivered to the USAF on November 30, 1970. 68-0060 is seen here in November 1972, wearing the red tail cap of the 77th TFS, 20th TFW. While assigned to the 20th TFW, 68-0060 crashed and was destroyed on November 5, 1975 near RAF Upper Heyford following a birdstrike. The bird penetrated the windscreen, causing loss of control. The crew, pilot Captain Steiber and WSO Captain Robert Gregory, ejected successfully, with one crew member receiving a major injury to his back. When the aircraft crashed, it had accumulated 407 flights and 1,204.3 flight hours. (Gerry Markgraf Collection)

68-0061 (A1-230) (E-71)
68-0061 was delivered to the USAF on November 25, 1970. Nicknamed *Big Dealer*, it spent its career assigned to the 20th TFW at Upper Heyford. It was a Desert Storm veteran and was named *Last Roll of the Dice* for its retirement, as seen here at AMARC on May 7, 1994. 68-0061 arrived at AMARC on December 8, 1993. When retired, the aircraft had a total of 5,707.4 flight hours. (Douglas Slowiak/Vortex Photo Graphics)

68-0062 (A1-231) (E-72)
68-0062 was delivered to the USAF on November 24, 1970. It is seen here on approach to Upper Heyford in December 1992. It was nicknamed *Land Shark*, and spent a majority of its career assigned to the 20th TFW at Upper Heyford. The aircraft arrived at AMARC on September 2, 1993. When retired, the aircraft had a total of 5,668.2 flight hours. (Jerry Geer Collection)

68-0063 (A1-232) (E-73)
68-0063 was delivered to the USAF on November 30, 1970. It is seen here in October 1985, wearing the yellow and black striped tail cap of the 79th TFS Tigers. Nicknamed *A Knight to Remember*, 68-0063 spent most of its career assigned to the 20th TFW at Upper Heyford. The aircraft, an AMP modified F-111E was reassigned to the 428th FS, 27th TFW on July 2, 1993. 68-0063 arrived at AMARC on December 12, 1995. When retired, the aircraft had a total of 6,085.3 flight hours. (Author's Collection)

68-0064 (A1-233) (E-74)
68-0064 was delivered to the USAF on November 30, 1970. It is seen here on August 1992, wearing the black tail cap of the 79th TFS, 20th TFW. It was named *6,000 General Dynamics Flyer*, after reaching 6,000 flight hours. 68-0064 spent its career assigned to the 20th TFW at Upper Heyford. The aircraft arrived at AMARC on November 17, 1993. When retired, the aircraft had a total of 6,171.8 flight hours. (Ben Knowles Collection)

68-0065 (A1-234) (E-75)
68-0065 was delivered to the USAF on December 29, 1970. It is seen here in November 1972, wearing the 77th TFS squadron tail code of JT. Nicknamed *The Armed Citizen*, 68-0065 spent its career assigned to the 20th TFW at Upper Heyford. 68-0065 arrived at AMARC on October 28, 1993. When retired, the aircraft had a total of 5,213.8 flight hours. (Gerry Markgraf Collection)

CHAPTER THREE: U.S. AIR FORCE TACTICAL F-111s

68-0066 (A1-235) (E-76)
The aircraft was delivered to the USAF on November 25, 1970, and is seen here on August 18, 1980, wearing the red tail cap of the 77th TFS, 20th TFW. 68-0066 was nicknamed *Crazy Horse* in 1982. While operating from Incirlik AB, Turkey, the 20th TFW aircraft crashed and was destroyed on July 20, 1990 at Sorgun, 62 miles from Incirlik. Smoke in the cockpit caused the crew to turn off the Yaw and Pitch Dampers, and the Central Air Data Computer System in an attempt to stop the smoke. The aircraft then stalled and departed controlled flight. The pilot, Captain R.W. Travis, and WSO, Captain R.M. Basak, ejected successfully. When lost, the aircraft had accumulated 2,048 flights and 5,346.9 flight hours. (Michael France)

68-0067 (A1-236) (E-77)
68-0067 was delivered to the USAF on December 9, 1970. It is seen here on July 16, 1979, and spent most of its career assigned to the 20th TFW at Upper Heyford. The aircraft, an AMP modified F-111E, was reassigned to the 428th FS, 27th FW. It arrived at AMARC on October 11, 1995. When retired, the aircraft had a total of 6,422.0 flight hours. (Geoff Rhodes)

68-0068 (A1-237) (E-78)
68-0068 was delivered to the USAF on November 30, 1970. It is seen here in January 1977, wearing the red tail cap of the 77th TFS, 20th TFW. It was nicknamed *One Man's Baby* in 1982 and *The Flak Ducker* in 1987-1988. 68-0068 spent most of its career assigned to the 20th TFW at Upper Heyford. The aircraft deployed to Incirlik, Turkey as part of Operation Proven Force, marked with 17 scores. 68-0068, an AMP modified F-111E was reassigned to the 428th FS, 27th FW on July 20, 1993. It arrived at AMARC on October 16, 1995. When retired, the aircraft had a total of 5,701.8 flight hours. (Michael France)

68-0069 (A1-238) (E-79)
68-0069 was delivered to the USAF on December 20, 1970. It is seen here at Nellis AFB in February 1988, wearing the red tail cap of the 77th TFS, and nose art *Wild Hare*. 68-0069 spent its career assigned to the 20th TFW at Upper Heyford. The aircraft deployed to Incirlik, Turkey as part of Operation Proven Force, and recorded 22 scores. It was renamed *Love Machine* in 1992. 68-0069 arrived at AMARC on November 17, 1993. When retired, the aircraft had a total of 5,940.9 flight hours. (Kevin Patrick)

68-0070 (A1-239) (E-80) NO PHOTO AVAILABLE
68-0070 was delivered to the USAF on December 30, 1970. The aircraft, assigned to the 20th TFW, crashed and was destroyed on October 31, 1977 near Welshpool England. The aircraft was flying night low level not using the TFR, when the aircraft hit the ground in a level 12 degree dive. The pilot Captain John J. Sweeney, and WSO, Captain William W. Smart, did not attempt ejection and were killed. The aircraft had accumulated 499 flights and 1,394.0 when it crashed.

68-0071 (A1-240) (E-81)
Above: 68-0071 was delivered to the USAF on December 28, 1970. It spent most of its career assigned to the 20th TFW at Upper Heyford. The aircraft, an AMP modified F-111E, was reassigned to the 428th FS, 27th FW on May 10, 1993. It is seen here on April 20, 1994 in the markings of the 428th FS at Cannon AFB. 68-0071 arrived at AMARC on May 16, 1996. When retired, the aircraft had a total of 5,26.8 flight hours. (Keith Snyder)

68-0072 (A1-241) (E-82)
Right: 68-0072 was delivered to the USAF on December 24, 1970. Nicknamed *Bad Medicine*, it is seen here at Nellis AFB during February, 1988 in 79th TFS markings. It spent most of its career assigned to the 20th TFW at Upper Heyford. 68-0072 deployed to Incirlik, Turkey as part of Operation Proven Force, and marked up 12 scores while flying missions in support of Desert Storm. 68-0072, an AMP modified F-111E was reassigned to the 428th FS, 27th FW. It arrived at AMARC on October 11, 1995. When retired, the aircraft had a total of 5,886.9 flight hours. (Kevin Patrick)

68-0073 (A1-242) (E-83)
68-0073 was delivered to the USAF on December 30, 1970. It is seen here marked as the 79th TFS Flagship when photographed in August 1986. The aircraft spent most of its career assigned to the 20th TFW at Upper Heyford. The aircraft, an AMP modified F-111E, was reassigned to the 428th FS, 27th FW. 68-0073 arrived at AMARC on October 16, 1995. When retired, the aircraft had a total of 5,530.0 flight hours. (Keith Snyder)

68-0074 (A1-243) (E-84)
68-0074 was delivered to the USAF on December 31, 1970. It is seen here at Nellis AFB in February 1988, wearing the white and black checkerboard tail cap of the 55th TFS, 20th TFW. The aircraft spent most of its career assigned to the 20th TFW at Upper Heyford. 68-0074 deployed to Incirlik, Turkey as part of Operation Proven Force, and recorded 30 scores. The aircraft, an AMP modified F-111E, was reassigned to the 428th FS, 27th FW on May 10, 1993. It's seen here at Cannon AFB on September 23, 1995. It was retired to AMARC on October 16, 1995, with a total of 5,870.3 flight hours. (Kevin Patrick)

68-0075 (A1-244) (E-85)
68-0075 was delivered to the USAF on December 31, 1970. Nicknamed Galleon, it spent most of its career assigned to the 20th TFW at Upper Heyford. 68-0075, an AMP modified F-111E, was reassigned to the 428th FS, 27th FW on May 10, 1993. It arrived at AMARC on December 13, 1995. When retired, the aircraft had a total of 6,070.2 flight hours. (Keith Snyder)

68-0076 (A1-245) (E-86)
68-0076 was delivered to the USAF on December 30, 1970. It is seen here in August 1983, wearing the blue tail cap of the 55th TFS, 20th TFW. The aircraft spent most of its career assigned to the 20th TFW at Upper Heyford. 68-0076 deployed to Incirlik, Turkey as part of Operation Proven Force, and recorded 28 scores. 68-0076, an AMP modified F-111E, was reassigned to the 428th FS, 27th FW on May 10, 1993, and retired to AMARC on December 15, 1995. (Scott Wilson)

68-0077 (A1-246) (E-87)
68-0077 was delivered to the USAF on December 31, 1970. It is seen here marked as the 77th TFS commanders aircraft, It was first named *June Nite*, and, in this photo, *Red Lady II*. The aircraft, an AMP modified F-111E, was reassigned to the 428th FS, 27th FW on July 20, 1993. 68-0077 arrived at AMARC on October 16, 1995. When retired, the aircraft had a total of 6,268.9 flight hours. (Author's Collection)

68-0078 (A1-247) (E-88)
68-0078 was delivered to the USAF on January 15, 1971. After spending most of its career assigned to the 20th TFW, 68-0078 was assigned to McClellan AFB, California for a time, wearing SM tail codes. While there, it was used as the kit proof aircraft for the Standard Flight Data Recorder (SFDR). It is seen here in the markings of the 428th FS. It was retired from the 27th Fighter Wing and arrived at AMARC on December 11, 1995. When retired, the aircraft had a total of 5,269.9 flight hours. (Alec Fushi)

CHAPTER THREE: U.S. AIR FORCE TACTICAL F-111s

68-0079 (A1-248) (E-89)
68-0079 was delivered to the USAF on January 31, 1971. Nicknamed *Tiger Lil* in 1987 and 1988, 68-0079 spent most of its career assigned to the 20th TFW at Upper Heyford. It deployed to Incirlik, Turkey as part of Operation Proven Force, marked with five Iraqi flags under the cockpit. 68-0079 seen here in September 1992, is marked as the 79th FS Flagship. The aircraft, an AMP modified F-111E, was reassigned to the 428th FS, 27th FW on May 10, 1993 with the nickname *Farewell Tiger*. 68-0079 arrived at AMARC on May 9, 1996. When retired, the aircraft had a total of 5,545.5 flight hours. (Douglas Slowiak Collection)

68-0080 (A1-249) (E-90)
The aircraft was delivered to the USAF on January 31, 1971. While assigned to the 20th TFW, the aircraft crashed on March 11, 1976. It was later repaired and returned to service. Nicknamed *Strange Brew*, 68-0080 spent most of its career assigned to the 20th TFW at Upper Heyford. The aircraft deployed to Incirlik, Turkey as part of Operation Proven Force, 68-0080 marked up 10 scores while flying missions in Desert Storm. The aircraft, an AMP modified F-111E, was reassigned to the 428th FS, 27th FW on May 10, 1993. 68-0080 is seen here at Cannon AFB on September 17, 1995. It retired to AMARC, arriving on December 11, 1995 with a total of 5,652.5 flight hours. (Pat Martin)

68-0081 (A1-250) (E-91) NO PHOTO AVAILABLE
The aircraft was delivered to the USAF on January 31, 1971. The aircraft, while assigned to the 20th TFW, crashed and was destroyed on March 5, 1975, near Leeming Bar, United Kingdom, following a birdstrike. The bird penetrated the windscreen, causing loss of control. The crew, pilot Major Wolfe and WSO Major Miller, ejected successfully with one crew member sustaining a back injury. The aircraft had accumulated 300 flights and 886.7 flight hours when it crashed.

68-0082 (A1-251) (E-92)
68-0082 was delivered to the USAF on January 31, 1971. On March 25, 1981, the aircraft suffered major damage during an aborted takeoff. The abort was caused by damage to an AOA probe which occurred during transient servicing. The nose gear was broken off and the forward fuselage, forward equipment bay, and the forward fuel tank were heavily damaged. The aircraft was airlifted by C-5A to General Dynamics' Restoration Facility, Fort Worth for repair. The forward fuselage of F-111A 65-5706 was modified to F-111E configuration and used to repair the aircraft. After the repair which cost $3.7 million, 68-0082 returned to operational service during October 1, 1982. During 1987 and 1988, 68-0082 was nicknamed *The Phoenix*. An AMP modified F-111E, 68-0082 was reassigned to the 428th FS, 27th FW. It is seen here at Cannon AFB during October 1994. 68-0082 was retired to AMARC on December 11, 1995, with a total of 5,535.5 flight hours. (Gerald McMasters)

68-0083 (A1-252) (E-93)
68-0083 was delivered to the USAF on January 31, 1971. It is seen here wearing the yellow and black striped tail cap of the 79th TFS, 20th TFW. Nicknamed *Prometheus II,* 68-0083 spent most of its career assigned to the 20th TFW at Upper Heyford. The aircraft deployed to Incirlik, Turkey as part of Operation Proven Force, it marked up 12 scores marked as Iraqi flags. 68-0083, an AMP modified F-111E, was reassigned to the 428th FS, 27th FW on July 20, 1993. It arrived at AMARC on December 11, 1995. When retired, the aircraft had a total of 6,066.5 flight hours. (Ben Knowles Collection)

68-0084 (A1-253) (E-94)
Right: 68-0084 was delivered to the USAF on February 28, 1971. It spent most of its operational career at Upper Heyford. It is seen here on July 20, 1986, in markings of the 55th TFS, 20th TFW. As an AMP modified F-111E, 68-0084 was reassigned to the 428th FS, 27th FW. Seen here at Cannon AFB in May 1995, the aircraft arrived at AMARC on December 13, 1995. When retired, the aircraft had a total of 6,032.2 flight hours. (Author's Collection)

F-111F

Procurement of stripped-down F-111Ds (already known as F-111Fs) was in the fiscal year 1970 budget which took effect on July 1, 1969. The avionics package (sometimes called the Mark IIF system) combined F-111D (excluding the AN/APN-189 Doppler Radar) and FB-111A navigation and digital computer systems, numerous other FB-111A components (such as the AN/APQ-144 attack radar), and some simpler, less costly avionics of earlier F-111s (the F-111E's stores management set included). The F-111F also featured an improved landing gear, the improved wing carry-through box, and the Pratt and Whitney new TF30-P-100 engine.

When the go-ahead decision was made on September 12, 1969, the USAF also disclosed that increased cost estimates forced it to limit production of the Mark II electronics and that future F-111s would have "a simpler and less costly system." Production approval came on June 19, 1970, several months after the aircraft's endorsement and for only 82 of 219 F-111Fs expected. Fifty-Eight were to be purchased in fiscal year (FY) 1970 and 24 in FY 1971. The USAF in mid-1960 wanted six F-111 Tactical Wings. This was cut to five in mid-1967 (one Wing of F-111As, one of F-111Es, and three of F-111Ds). In 1969, the three F-111D Wings were reduced to one, with the remaining two Wings due to be equipped with cheaper F-111Fs. At year-end, another money-saving change cut the two F-111F Wings to one.

With the thrust of the F-111D's P-9 engine still too low, the USAF, in September 1968, ordered development of the still more powerful P-100 engine. This engine was first planned to be installed in the 107th F-111D. Even though the thrust of the P-9 surpassed that of the P-3 of the F-111A and F-111E aircraft, it did not give the F-111D all the maneuverability the USAF wanted. The P-100 was the sixth in the Pratt & Whitney TF30-P series of turbofan engines which had been installed in different versions of the F-111. It was decided in September 1969, (when the F-111D program was cut to 96 aircraft) that the P-100 would equip the F-111Fs.

The TF30-P-100 engine generated 25,100 pounds of thrust with afterburner. This was 5,500 more pounds than the P-9, boosting takeoff thrust by 40 percent. To reduce drag, an adjustable nozzle buried in the engine exhaust section was utilized. Between January and March 1971, the P-100 was tested on an F-111A. Engine and airframe were compatible, which reduced the engine's Category I flight tests by almost 40 percent. The ground tests did not go well (the engine failed after 147 hours), but the three engineering changes required to fix the problem were not expected to affect the engine delivery schedule. On June 18, 1971, a turbine blade broke on a P-100 production engine during checkout at the General Dynamics, Fort Worth plant. As a result, early F-111Fs, due for delivery beginning in September were equipped with P-9 engines. The USAF thought only 31 F-111Fs would be involved, but additional technical problems slipped delivery of the new P-100 engines to the spring of 1972. By then, the USAF had accepted 49 P-9-equipped F-111Fs. These were retrofitted with P-100 engines as soon as possible. The re-engining task was completed at General Dynamics, Fort Worth on July 3, 1972.

The F-111F entered operational service with the 347th Tactical Fighter Wing at Mountain Home AFB, Idaho on September 20, 1971. Initially, the F-111Fs were scheduled to be assigned to the 31st TFW at Homestead AFB, Florida. However, TAC's request to send the aircraft to the 347th was approved on December 3, 1970.

One squadron of the 347th TFW reached Initial Operational Capability (IOC) a few months after the F-111F entered operational service. The entire wing became operationally ready in October 1972, one month ahead of schedule.

The significant F-111F difficulties stemmed from the P-100 engine. With the onset of cold weather at Mountain Home, afterburner stalls, one of several problems believed to be solved, re-occurred. Modification of the engines was completed by November 11, 1972. A plastic diaphragm in the afterburner turn-on switch, operating poorly in low temperatures, had caused the problem and was replaced. Several other engine deficiencies (tail-feather seal leakage, inlet guide vane cracking, etc.) were also corrected by the end of 1972. The inspection of two P-100 engines with 300 hours of flight time had disclosed an accumulation of atmospheric dust in the

The four F-111Fs from the 27th Fighter Wing seen here attended the formal retirement and naming ceremony of the F-111 Aardvark held at Fort Worth on July 27, 1996. (Lockheed Martin Tactical Aircraft Systems)

The lineup of 48th TFW F-111Fs occurred in March 1981 for the RAF Bombing Competition at RAF Lossie Mouth. The number 1 aircraft, not in the photo was 72-1443 (Chris McWilliams Collection)

engine's blade cavity. The dust did not harm the engine's life or its operation for 450 hours. However, it did damage the second turbine inner air seal. This was fixed by the installation of a new blade with a drilled hole in its tip, which allowed the dust escape. By June 30, 1973 the P-100's operational life had risen to 600 hours. One month later Pratt & Whitney indicated that the time between overhauls (always too short for hard-to-get new engines) could be extended to about 2,000 hours by reducing the P-100's maximum thrust to 23,000 pounds. The 23,000 pounds were still 4,500 pounds above the P-3 of the F-111A, F-111E, and F-111C; 3,400 pounds above the P-9 of the F-111D; and 2,750 above the P-7 of the FB-111A. Remaining problems and improvements awaited a forthcoming engine's update program.

The 347th wing at Mountain Home had fewer supply problems with its F-111Fs than did the 27th Wing at Cannon with more complex F-111Ds. Additionally, the operational rate at Mountain Home exceeded that of the longer established F-111A and F-111D Wings.

The last of the F-111Fs was originally scheduled for delivery in December 1974. Production was extended until completion of all F-111Fs, set for late 1976. Seventy-six of the programmed 94 F-111Fs were accepted by December 1974. The total was raised to 106 due to Congress' desire to keep the production line open, even though the Department of Defense did not want to release more F-111 money. The USAF accepted 70 F-111Fs in fiscal year (FY) 1972; none were delivered during the first half of FY 1973. Deliveries resumed in January 1973 at a monthly rate of one aircraft until the 106th aircraft was delivered.

The flyaway cost per aircraft was $10.3 million; divided as follows: airframe – $5,097,000; engines (installed) – $2,026,000; electronics – $1,711,000; ordnance – $6,000; and armament – $1,529,000. A post-FY 73 cost increase of the F-111F airframe raised the aircraft unit price to $10.9 million. Added to the RDT&E costs, this gave the F-111F a price tag of $13.7 million.

The 366th TFW replaced the 347th TFW at Mountain Home AFB on October 30, 1972, and assumed control of the former 347th TFW F-111F assets, including the 389th TFS, 390th TFS, and 391st TFS.

On August 16, 1976, as a result of a U.S. Army officer being killed by the North Koreans, in what came to be called the Tree Cutting Incident, F-111Fs of the 366th TFW deployed to South Korea. The following 19 jets deployed with just 12 hours notice: 70-2362, 70-2363, 70-2380, 70-2383, 70-2385, 70-2390, 70-2391, 70-2396, 70-2404, 70-2406, 70-2408, 70-2416, 71-0890, 72-1442, 72-1447, 72-1451, 74-0177, 74-0180, and 74-0183.

After one mid-1970s NATO exercise in which bad weather had grounded all flights except Upper Heyford's F-111Es, it was decided NATO needed more F-111s. The result was Operation Creek Swing/Ready Switch, which transferred the F-111Fs to RAF Lakenheath, England. Under Operation Creek Swing/Ready Switch in July and August 1977, the 366th converted from F-111Fs to F-111As. The F-111Fs of the 366th TFW transferred to the 48th TFW at RAF Lakenheath. The 48th TFW consisted of three Tactical Fighter Squadrons; the 492nd TFS, 493rd TFS, and the 494th TFS. A fourth squadron, the 495th TFS activated within the 48th TFW in 1977 to provide difference training for aircrews coming from other F-111 models.

As part of Operation Desert Shield to protect Saudi Arabia, the 48th TFW deployed from Lakenheath to Taif, Saudi Arabia on August 25, 1980, increasing numbers until a total of 67 F-111Fs were in place on January 17, 1991, at the start of Operation Desert Storm. (See Desert Shield/Desert Storm in the COMBAT OPERATIONS section).

The last elements of 48th TFW returned from Taif to Lakenheath during May 1991. The F-111Fs were transferred to the 27th TFW at Cannon during 1992 as they were replaced at Lakenheath with F-15s. The 492nd and 494th converted to F-15Es in 1992. The 493rd and 495th were inactivated in December 1992. The 493rd was activated with F-15C/D s on January 1, 1994.

The F-111F's avionics were updated under the Pacer Strike program, receiving similar equipment to that of the AMP F-111Es.

The F-111Fs operated at Cannon, assigned first to the 524th TFS, followed by the 522nd TFS, and 523rd TFS, until retired to AMARC during 1996. The 524th TFS was equipped with the F models modified to carry the GBU-15 and its rocket powered brother, the AGM-130. The last operational F-111F, 74-0187, arrived at AMARC on July 29, 1996, along with 71-0888 and 74-0178.

F-111F

Most of the 106 F-111Fs started their careers assigned to the 347th TFW at Mountain Home AFB, Idaho. The F-111F factory numbers used the E2 series starting with E2-01 and continuing through E2-106. The 347th TFW was replaced by the 366th TFW on October 30, 1972, with the F-111F transferring, in place, to the new TFW. As a result of the Tree Cutting Incident by North Korea, 19 F-111Fs deployed to South Korean as a show of force. Under operation Ready Switch 98 F-111Fs TFW were transferred from the 366th TFW to the 48th TFW at RAF Lakenheath Upper Heyford, England. From Lakenheath, F-111Fs took part in Operation El Dorado Canyon, the attack on Libya. Sixty-seven F-111Fs deployed to Taif AB, Saudi Arabia flying missions during 1990 and 1991 in support of Operation Desert Shield and Desert Storm. After deployment to Taif, the aircraft returned to Lakenheath. The F-111Fs of the 48th TFW were replaced with F-15s beginning in 1992. Like the Upper Heyford F-111Es, the F-111Fs deployed back to the U.S. going to Cannon AFB, New Mexico. The F-111Fs were retired to AMARC during 1995 and 1996.

70-2362 (E2-01) (F-01)
The aircraft first flew in August 1971. 70-2362 was delivered to the USAF on October 14, 1971. It was Pave Tack modified and flew in Desert Storm. Also receiving the Pacer Strike Modification, it was last assigned to the 524th FS at Cannon, as seen here on September 23, 1995. 70-2362 was retired to AMARC on July 29, 1996. When retired, the aircraft had a total of 5,598.2 flight hours. (Keith Snyder)

70-2363 (E2-02) (F-02)
70-2363 was delivered to the USAF on September 16, 1971. It is seen here in July 1982. It was Pave Tack modified and flew in Operation El Dorado Canyon, callsign ELTON 43. It also flew in Desert Storm. 70-2363 aircraft arrived at AMARC on June 6, 1996. When the aircraft retired, it had a total of 6,113.6 flight hours. (Jim Rotramel)

70-2364 (E2-03) (F-03)
70-2364 was delivered to the USAF on September 16, 1971. It is seen here at McClellan AFB on October 18, 1974, with the red tail stripe of the 389th TFS, 366th TFW. The aircraft was Pave Tack modified and flew in Desert Storm. After assigned to Cannon AFB, 70-2364 was named "The City of Portales" and during 1996, was put on display on a pylon on the south side of Portales, New Mexico, in the median of US 70. (Author)

70-2365 (E2-04) (F-04)
70-2365 was delivered to the USAF on September 10, 1971. The aircraft was Pave Tack modified aircraft and flew in Desert Storm. 70-2365 is seen here at Nellis AFB on October 19, 1993 marked as the 524th FS Flagship. It arrived at AMARC on November 13, 1995. When retired, the aircraft had a total of 6,131.0 flight hours. (Brian C. Rogers)

CHAPTER THREE: U.S. AIR FORCE TACTICAL F-111s

70-2366 (E2-05) (F-05)
70-2366 was delivered to the USAF on October 1, 1971. Seen here on March 20, 1976, it was marked as the 366th TFW's Commanders Flagship. (Author's Collection)

70-2366, assigned to the 493rd TFS, 48th TFW, is seen here over Turkey during December 1982. While assigned to the 48th TFW, the aircraft crashed and was destroyed on December 21, 1983, in the North Sea, near Scarborough UK. The aircraft departed controlled flight after leaving a tanker following a night refueling. The crew ejected safely. When the aircraft was lost, it had accumulated 1,180 flights and 3,109.6 flight hours. (Jim Rotramel)

70-2367 (E2-06) (F-06) NO PHOTO AVAILABLE
The aircraft was delivered to the USAF on October 14, 1971. On April 20, 1979, while assigned to the 48th TFW, 70-2367 crashed into the Donoch Firth and was destroyed, This was as a result of a low level mid-air with 73-0714, in the Tain Ranges, Scotland. The crew successfully ejected. 70-2367 had accumulated 637 flights and 1,736.0 flight hours when it was lost.

70-2368 (E2-07) (F-07)
The aircraft was delivered to the USAF on October 14, 1971. 70-2368 is seen here on December 3, 1983. It was Pave Tack modified. While assigned to the 492nd TFS, 48th TFW, 70-2368 crashed and was destroyed on May 2, 1990, over Sculthorpe, England. While inbound to RAF Sculthorpe, the crew of REX 62 heard a loud thump. A first stage fan failure had caused the noise and resultant fire. The crew ejected successfully. The aircraft crashed in a forest, two miles from RAF Sculthorpe. At the time of the crash, the aircraft had accumulated 1,692 flights and 4,269.8 flight hours. (Scott Wilson)

70-2369 (E2-08) (F-08)
70-2369 was delivered to the USAF on October 21, 1971. It was Pave Tack modified and flew in Desert Storm. 70-2369 is seen here marked as the 522nd FS Flagship. It arrived at AMARC on January 11, 1996. When retired, the aircraft had a total of 5,887.6 flight hours. (Keith Snyder)

70-2370 (E2-09) (F-09)
70-2370 was delivered to the USAF on October 14, 1971. It was Pave Tack modified and flew in Desert Storm. 70-2370 is seen here, while assigned to the 524th FS, releasing a GBU-15(V)31. It was written off in July 1994, after a weapons bay fire, while at Nellis AFB. All useful parts were removed, and the remainder of the aircraft was scrapped. (USAF via Marty Isham)

70-2371 (E2-10) (F-10)
70-2371 was delivered to the USAF on October 23, 1971. It was Pave Tack modified and flew in Operation El Dorado Canyon, callsign JEWEL 61. It also flew in Desert Storm. It is seen here at Langley AFB on April 13, 1993, marked as the 523rd FS Flagship. After being assigned to the 27th FW at Cannon AFB, 70-2371 was retired to AMARC on January 11, 1996. When retired, the aircraft had a total of 5,959.2 flight hours. (Brian C. Rogers)

70-2372 (E2-11) (F-11)
70-2372 was delivered to the USAF on October 22, 1971. It is seen here in August 1974, with the red tail stripe of the 389th TFS, 366th TFW. The aircraft was Pave Tack modified. 70-2372 arrived at AMARC on January 8, 1996. When the aircraft retired, it had a total of 5,948.3 flight hours. (Tom Brewer Collection)

70-2373 (E2-12) (F-12)
70-2373 was delivered to the USAF on November 1, 1971. It is seen here in July 1987, approaching refueling position. It was assigned to the 492nd TFS, 48th TFW at the time the photo was taken. The aircraft was Pave Tack modified. 70-2373 arrived at AMARC on October 18, 1995. When the aircraft retired it, a had total of 5,672.3 flight hours. (Jim Rotramel)

70-2374 (E2-13) (F-13)
70-2374 was delivered to the USAF on November 9, 1971. It is seen here in September 1987, wearing the yellow tail cap of the 493rd TFS, 48th TFW. It was Pave Tack modified and arrived at AMARC on October 18, 1995. When the aircraft retired, it had a total of 5,880.0 flight hours. (Kevin Foy)

70-2375 (E2-14) (F-14)
70-2375 was delivered to the USAF on November 8, 1971. 70-2375 is seen here in October 1986, carrying GBU-12B/B 500 pound laser guided bombs. It's marked with the yellow tail cap of the 493rd TFS, 48th TFW. While assigned to the 493rd TFS, 48th TFW, 70-2375 crashed and was destroyed on July 28, 1987, in Scotland. The aircraft entered the weather (Instrument Meteorological Conditions – IMC) during a day toss maneuver. It exited the weather (IMC) nose low, too low to recover, and hit the ground at over 500 knots. Both air crew members, Captains Thomas "Chip" Stem and Phillip "Phil" Baldwin, were killed. When the aircraft was lost, it had accumulated 1,938 flights and 4,939.6 flight hours. (Jim Rotramel)

70-2376 (E2-15) (F-15)
70-2376 was delivered to the USAF on November 1, 1971. It was painted as the 366th TFW's Bicentennial Aircraft as seen here in this photo taken on August 7, 1976. The aircraft was Pave Tack modified. 70-2376 arrived at AMARC on January 9, 1996. When the aircraft retired, it had a total of 6,270.5 flight hours. (Stephen Miller)

70-2377 (E2-16) (F-16)
70-2377 was delivered to the USAF on November 17, 1971. It is seen here on a snowy ramp at Offutt AFB in February 1982 marked with the yellow tail cap of the 493rd TFS, 48th TFW. It was a Pave Tack modified aircraft. While assigned to the 48th TFW, the aircraft crashed and was destroyed on December 7, 1982 off the Isle of Skye Scotland. The aircraft initiated a Auto TF Descent over water, and hit a cliff during level off. Ejection was not attempted, and both crew members, Major Burnley L. Rudiger and 1 Lt Steven J. Pitt, were killed. When the aircraft crashed, it had accumulated 941 flights and 4,153.9 flight hours. (George Cockle)

CHAPTER THREE: U.S. AIR FORCE TACTICAL F-111s

70-2378 (E2-17) (F-17)
70-2378 was delivered to the USAF on November 8, 1971. It was badly damaged on May 1, 1979 following an engine failure and inflight fire which burned into the saddle fuel tank. The aircraft landed safely, but had major damage to the left and right stabilators, the vertical fin, and both saddle tanks. It was repaired and returned to operational service. 70-2378 was Pave Tack modified and flew in Desert Storm. It is seen here on April 20, 1994 at Cannon AFB, marked with the blue tail stripe of the 523rd FS. 70-2378 arrived at AMARC on March 25, 1996. When retired, the aircraft had a total of 4,642.9 flight hours. (Keith Snyder)

70-2379 (E2-18) (F-18)
70-2379 was delivered to the USAF on November 11, 1971. It is seen here at Aviano AB, Italy with the yellow tail cap of the 493rd TFS, 48th TFW. The aircraft was Pave Tack modified and flew in Desert Storm. 70-2379 was later assigned to Eglin AFB, and arrived at AMARC on October 18, 1995. When retired, the aircraft had a total of 5322.1 flight hours. (Author's Collection)

70-2380 (E2-19) (F-19) NO PHOTO AVAILABLE
70-2380 was delivered to the USAF on November 11, 1971. While assigned to the 48th TFW, the aircraft crashed and was destroyed on December 15, 1977. During a Functional Check Flight, after declaring an Inflight Emergency (IFE) for a utility hydraulic system failure, the aircraft experienced an uncontrolled pitch up. The crew ejected and was not injured. The aircraft crashed near a school in Newmarket UK. The aircraft had accumulated 550 flights and 1,542.5 flight hours at the time of the loss.

70-2381 (E2-20) (F-20)
70-2381 was delivered to the USAF on December 1, 1971. Seen here in November 1983, 70-2381 awaits while its wingman receives fuel from KC-135A 62-3521. It carries the yellow tail cap of the 493rd TFS. 70-2381 was Pave Tack modified and arrived at AMARC on October 19, 1995. When the aircraft retired, it had a total of 5,920.7 flight hours. (Jim Rotramel)

70-2382 (E2-21) (F-21)
70-2382 was delivered to the USAF on December 1, 1971. It is seen here in July 1978, with the yellow tail cap of the 493rd TFS, 48th TFW. The aircraft was Pave Tack modified and flew in Operation El Dorado Canyon, callsign REMIT 34. 70-2382 arrived at AMARC on January 4, 1996. When retired, the aircraft had a total of 5,160.8 flight hours. (Terry Love Collection)

70-2383 (E2-22) (F-22)
70-2383 was delivered to the USAF on December 22, 1971. It was Pave Tack modified and flew in Operation El Dorado Canyon, callsign JEWEL 62, It also flew in Desert Storm. 70-2383 seen here in March 1993, is marked as the 27th FW Operations Group (OG) Commanders Flagship. It arrived at AMARC on October 26, 1995. When the aircraft was retired, it had a total of 6,057.1 flight hours. (Douglas Slowiak/Vortex Photo Graphics)

CHAPTER THREE: U.S. AIR FORCE TACTICAL F-111s

70-2384 (E2-23) (F-23)
70-2384 was delivered to the USAF on December 8, 1971. It was Pave Tack modified and flew in Desert Storm. It was damaged in a collision with a KC-135 tanker on January 17, 1991, the first night of Desert Storm. 70-2384 is seen here on October 14, 1993, with the yellow tail stripe of the 524th FS, 27th FW. 70-2384 arrived at AMARC on January 9, 1996. When retired, the aircraft had a total of 5,886.8 flight hours. (Brian C. Rogers)

70-2385 (E2-24) (F-24)
70-2385 was delivered to the USAF on November 24, 1971. It is seen here in August 27, 1983, with the green tail cap of the 495th TFS, 48th TFW. The aircraft was Pave Tack modified. 70-2385 arrived at AMARC on November 6, 1995. When retired, the aircraft had a total of 5,838.1 flight hours. (Scott Wilson)

70-2386 (E2-25) (F-25)
70-2386 was delivered to the USAF on December 28, 1971. It is seen here in July 1979, with the yellow tail cap of the 493rd TFS, 48th TFW. The aircraft was Pave Tack modified and flew in Operation El Dorado Canyon, callsign JEWEL 64. 70-2386 also flew in Desert Storm. It arrived at AMARC on January 8, 1996. When retired, the aircraft had a total of 6,277.7 flight hours. (Scott Wilson)

70-2387 (E2-26) (F-26)
70-2387 was delivered to the USAF on January 13, 1972. It is seen here in August 1981, with the yellow tail cap of the 493rd TFS, 48th TFW. The aircraft was Pave Tack modified and flew in Operation El Dorado Canyon, callsign LUJAC 23. It also flew in Desert Storm. 70-2387 arrived at AMARC on December 4, 1995. When retired, the aircraft had a total of 5,947.2 flight hours. (Terry Love Collection)

70-2388 (E2-27) (F-27)
Right: 70-2388 was delivered to the USAF on December 28, 1971. It is seen here on September 15, 1972, with the blue tail stripe of the 391st TFS, 366th TFW. While assigned to the 366th TFW at Mountain Home AFB, the aircraft crashed and was destroyed on March 16, 1976. The aircraft crashed on the runway during landing when a throttle stuck in military power (100% engine power – but no afterburner). The crew ejected successfully with no injuries. When the aircraft was lost, it had accumulated 386 flights and 1,112.2 flight hours. (Steven Miller)

70-2389 (E2-28) (F-28)
70-2389 was delivered to the USAF on January 12, 1972. It is seen here in May 1980, with the yellow tail cap of the 493rd TFS, 48th TFW. It was Pave Tack modified. While assigned to the 48th TFW, the aircraft was destroyed during the Libyan Raid (Operation El Dorado Canyon) on April 15, 1986. This F-111 was crewed by Major Fernando Ribas Dominici (pilot) and Capt. Paul Lorence (WSO). Pilots who flew on the mission reported seeing a fireball fall into the sea during the raid. In the light of a lack of any other evidence, the USAF concluded that the plane was lost to a SAM or AAA. After years of denying that they had the bodies of the two crew members, the Libyan authorities returned the remains of Fernando on request from the Pope. He was identified by dental records. Paul Lorence has never been returned. The callsign was KARMA 52. When the aircraft was lost, it had accumulated 1,286 flights and 3,294.5 flight hours. (G. Salerno)

70-2390 (E2-29) (F-29)
70-2390 was delivered to the USAF on December 28, 1971. It was Pave Tack modified, flying in Operation El Dorado Canyon on April 14, 1986 as REMIT 31, the lead aircraft for the attack on Al Azziziyah Barracks Compound. It was also the lead F-111F deployed to Taif, Saudi Arabia, on the first day of Operation Desert Shield August 1990. It completed 29 combat sorties during Desert Storm, destroying 58 targets. 70-2390 is seen here in June 1991, marked as the 48th TFW Flagship. It is carrying GBU-24A Paveway III 2000 pound low level laser guided bombs. Upon retirement, it was flown to USAF Museum Wright-Patterson by Lt. Col. Bob Brewster and Maj. Fred Cheney on May 14, 1996. It used callsign KARMA 52 in respect to the crew of F-111F 70-2389 lost during Operation El Dorado Canyon. (Ben Knowles Collection)

70-2391 (E2-30) (F-30)
70-2391 was delivered to the USAF on December 22, 1971. It was Pave Tack modified and flew in Desert Storm. On February 27, 1991, 70-2391 was used to deliver the first operational GBU-28 5000 lb class Pave Way III laser guided bomb. The bomb is made from a 155 inch howitzer barrel filled with Tritonal explosive, to which a GBU-27 Pave Way II kit is attached. The target was Al Taji Air Base in Iraq. 70-2391 is seen here on March 15, 1994, with the blue tail stripe of the 523rd FS, 27th FW, and special tail markings outlined in white. It arrived at AMARC on April 8, 1996. When retired, the aircraft had a total of 6,221.9 flight hours. (Brian C. Rogers)

70-2392 (E2-31) (F-31)
Left: 70-2392 was delivered to the USAF on January 12, 1972. It was Pave Tack modified and flew in Operation El Dorado Canyon, callsign PUFFY 11. 70-2392 was also involved in Operation Desert Storm. It is seen here at Nellis AFB on March 5, 1976, with the blue tail stripe of the 391st TFS, 366th FW. It arrived at AMARC on October 26, 1995. When retired, the aircraft had a total of 5,997.5 flight hours. (Author)

70-2393 (E2-32) (F-32) NO PHOTO AVAILABLE
70-2393 was delivered to the USAF on January 27, 1972. While assigned to the 366th TFW at Mountain Home AFB, 70-2393 crashed and was destroyed on November 8, 1975, after the aircraft lost control during low speed flight. The crew ejected successfully with one crew member receiving back injuries. When the aircraft was lost, it had accumulated 325 flights and 906.4 flight hours.

70-2394 (E2-33) (F-33)
70-2394 was delivered to the USAF on January 26, 1972. It is seen here in May 1981, with the yellow tail cap of the 493rd TFS, 48th TFW. It was Pave Tack modified and flew in Operation El Dorado Canyon, callsign PUFFY 13. It also flew in Desert Storm. It arrived at AMARC on April 8, 1996. When retired, the aircraft had a total of 5,977.4 flight hours. (Terry Love Collection)

70-2395 (E2-34) (F-34) NO PHOTO AVAILABLE
70-2395 was delivered as the 400th F-111 on January 21, 1972. 70-2395 crashed and was destroyed on September 11, 1974, at Saylor Creek Range while assigned to the 366th TFW at Mountain Home AFB. The aircraft crashed short of the target while on a night bombing run. Ejection was not attempted, Captain William R. Kennedy and Captain David C. McKennon were killed. The aircraft had accumulated 264 flights and 744.2 flight hours when it crashed.

70-2396 (E2-35) (F-35)
70-2369 was delivered to the USAF on February 14, 1972. It is seen here on September 15, 1972, with the blue tail stripe of the 391st TFS. The 347th TFW emblem can be seen behind the cockpit. The aircraft was Pave Tack modified and flew in Operation El Dorado Canyon, callsign ELTON 42. 70-2396 also flew in Desert Storm. It arrived at AMARC on January 3, 1996. When retired, the aircraft had a total of 6,118.3 flight hours. (Marty Isham Collection)

CHAPTER THREE: U.S. AIR FORCE TACTICAL F-111s

70-2397 (E2-36) (F-36)
The aircraft was delivered to the USAF on January 25, 1972, and was Pave Tack modified. It was one of the first three delivered to RAF Lakenheath as part of Operation Ready Switch. 70-2397 is seen here on July 16, 1983, with the yellow tail cap of the 493rd TFS, 48th TFW. It crashed and was destroyed on April 5, 1989 while assigned to the 494th TFS, 48th TFW. It was flying with the callsign GREEBIE 54 on a RED FLAG mission over the Nellis Range complex. The aircraft slowed to below 200 knots while crossing ridge lines. The aircraft impacted at the crest of one of the ridges, killing the crew, 1 Lt Bob Boland and Captain James Gleason. When the aircraft crashed, it had accumulated 1,526 flights and 3,868.0 flight hours. (Scott Wilson)

70-2398 (E2-37) (F-37)
70-2398 was delivered to the USAF on February 9, 1972. It was Pave Tack modified and flew in Desert Storm. It's seen here at Langley AFB on May 6, 1993 in the markings of the 27th FW Flagship. It is carrying MK 82 AIR 500 pound bombs with a Pave Tack pod in the weapons bay. 70-2398 arrived at AMARC on January 9, 1996. When retired, the aircraft had a total of 5,945.2 flight hours. (Brian C. Rogers)

70-2399 (E2-38) (F-38)
70-2399 was delivered to the USAF on February 11, 1972. It was Pave Tack modified and flew in Desert Storm. 70-2389 is seen here on June 4, 1995, with the yellow tail stripe of the 524th FS, 27th FW. It arrived at AMARC on July 8, 1996. When retired, the aircraft had a total of 5,664.2 flight hours. (Brian C. Rogers)

70-2400 (E2-39) (F-39)
70-2400 was delivered to the USAF on February 9, 1972. It was Pave Tack modified. 70-2400 is seen here on March 22, 1995, with the ET tail code of the 40th Flight Test Squadron, 46th Test Wing, Eglin AFB. The orange blocks on the fuselage and wing tips are camera mounts used for weapon release recording cameras. 70-0400 arrived at AMARC on October 19, 1995. When retired, the aircraft had a total of 4,228.8 flight hours. (Tony Sacketos)

70-2401 (E2-40) (F-40)
70-2401 was delivered to the USAF on February 24, 1972. It is seen here on March 17, 1980, with the yellow tail cap of the 493rd TFS, 48th TFW. It was Pave Tack modified and flew in Desert Storm. It arrived at AMARC on January 9, 1996. When retired, the aircraft had a total of 5,917.3 flight hours. (Douglas Slowiak Collection)

70-2402 (E2-41) (F-41)
70-2402 was delivered to the USAF on March 6, 1972. It is seen here at Langley AFB on August 30, 1994, with the red tail stripe of the 522nd FS, 27th FW. It was Pave Tack modified and flew in Desert Storm. It arrived at AMARC on January 8, 1996. When retired, the aircraft had a total of 5,992.1 flight hours. (Brian C. Rogers)

70-2403 (E2-42) (F-42)
70-2403 was delivered to the USAF on February 23, 1972. It is seen here at Nellis in May 1975, with the blue tail stripe of the 391st TFS, 366th TFW. It was Pave Tack modified and flew in Operation El Dorado Canyon, callsign ELTON 41. 70-2403 also flew in Desert Storm. It arrived at AMARC on April 4, 1996. When retired, the aircraft had total of 5,645.9 flight hours. (Author)

70-2404 (E2-43) (F-43)
70-2404 was delivered to the USAF on February 28, 1972. It is seen here in April 20, 1994, with the blue tail stripe of the 523rd FS, 27th FW, and special tail markings outlined in white. It was Pave Tack modified and flew in Operation El Dorado Canyon, callsign ELTON 44. It also flew in Desert Storm. 70-2404 arrived at AMARC on October 26, 1995. When retired, the aircraft had a total of 5,987.0 flight hours. (Keith Snyder)

70-2405 (E2-44) (F-44)
70-2405 was delivered to the USAF on February 28, 1972. It is seen here with the yellow tail cap of the 493rd TFS, 48th TFW. It was Pave Tack modified and flew in Operation El Dorado Canyon, callsign LUJAC 24. It also flew in Desert Storm. 70-2405 received the first Pacer Strike Modification and flew with the 524th FS at Cannon AFB. It arrived at AMARC on April 2, 1996. When retired, the aircraft had a total of 5,017.7 flight hours. (Terry Love Collection)

70-2406 (E2-45) (F-45)
Above: 70-2406 was delivered to the USAF on March 26, 1972. It is seen here on November 11, 1972, with the blue tail stripe of the 391st TFS, 347th TFW. The aircraft was Pave Tack modified and flew in Desert Storm. 70-2406 arrived at AMARC on March 22, 1996. When retired, the aircraft had a total of 6,021.0 flight hours. (Tom Brewer)

70-2407 (E2-46) (F-46) NO PHOTO AVAILABLE
70-2407 was delivered to the USAF on March 15, 1972. It was destroyed as a result of a massive engine fire at engine run station #3 on February 2, 1972, before delivery from General Dynamics. The aircraft was accepted by the USAF and the DD 250 signed on March 15, 1972 after an investigation of the cause. The aft fuselage section was used to repair F-111D 68-0127.

70-2408 (E2-47) (F-47)
Right: 70-2408 was delivered to the USAF on April 14, 1972. It was Pave Tack modified and flew in Desert Storm. It's seen here at Nellis AFB on October 19, 1993 marked as the 522nd FS Flagship. 70-2408 is now on display in Santa Fe, New Mexico. (Brian C. Rogers)

70-2409 (E2-48) (F-48)
70-2409 was delivered to the USAF on March 13, 1972. It was Pave Tack modified and flew in Desert Storm. 70-2409 is seen here at Cannon AFB in May 1995, with the blue tail stripe of the 523rd FS, 27th FW. It arrived at AMARC on April 4, 1996. When retired, the aircraft had a total of 6,070.7 flight hours. (Alec Fushi)

CHAPTER THREE: U.S. AIR FORCE TACTICAL F-111s

70-2410 (E2-49) (F-49) NO PHOTO AVAILABLE
Delivered to the USAF on March 24, 1972, 70-2410 crashed and was destroyed three months later. While assigned to the 347th TFW at Mountain Home AFB, the aircraft crashed on June 15, 1972, near Fallon, Nevada. The aircraft departed controlled flight at 15,000 feet while increasing distance from the formation lead. The crew successfully ejected at 7,000 feet, but one crew member received a back injury on landing. When the aircraft crashed, it was almost brand new, having accumulated only 24 flights and 65.6 flight hours.

70-2411 (E2-50) (F-50)
70-2411 was delivered to the USAF on March 14, 1972. It was Pave Tack modified and flew in Desert Storm. 70-2411 is seen here on May 22, 1994, with the yellow tail stripe of the 524th FS, 27th FW. It arrived at AMARC on July 10, 1996. When retired, the aircraft had a total of 5,765.9 flight hours. (Norris Graser)

70-2412 (E2-51) (F-51)
70-2412 was delivered to the USAF on March 23,1972. It is seen here on July 23, 1993, with the red tail stripe of the 522nd FS, 27th FW. It was Pave Tack modified. 70-2389 is seen here in May 1980, with the yellow tail cap of the 493rd TFS, 48th TFW. While assigned to the 522nd FS, 27 FW, Cannon AFB, this aircraft crashed and was destroyed on September 22, 1993 on the Melrose Ranges. The aircraft had a right engine failure over the range. The aircraft was turned back to Cannon, but the resulting fire started to affect control of the aircraft, so the pilot zoomed the aircraft to 12,000 feet. The pilot, Major Robby Kyorauc, and WSO, Captain Gregory Wilson, ejected successfully. (Norris Graser)

70-2413 (E2-52) (F-52)
70-2413 was delivered to the USAF on April 4, 1972. It was Pave Tack modified and flew in Operation El Dorado Canyon, callsign KARMA 51. 70-2413 also flew in Desert Storm. It is seen here in May 1995, with the red tail stripe of the 522nd FS, 27th FW. It arrived at AMARC on January 8, 1996. When retired, the aircraft had a total of 5,687.2 flight hours. (Alec Fushi)

70-2414 (E2-53) (F-53)
70-2414 was delivered to the USAF on March 29, 1972. It was Pave Tack modified and flew in Desert Storm. 70-2414 is seen here on October 24, 1993, with the blue tail stripe of the 523rd FS, 27th FW. It arrived at AMARC on April 8, 1996. When retired, the aircraft had a total of 5,695.2 flight hours. (Keith Snyder)

70-2415 (E2-54) (F-54)
702415 was delivered to the USAF on April 18, 1972. It was Pave Tack modified and flew in Operation El Dorado Canyon, callsign KARMA 54, and also flew in Desert Storm. 70-2415 arrived at AMARC on October 19, 1995. When retired, the aircraft had a total of 5,949.9 flight hours. (USAF)

CHAPTER THREE: U.S. AIR FORCE TACTICAL F-111s

70-2416 (E2-55) (F-55)
70-2416 was delivered to the USAF on April 18, 1972. It was Pave Tack modified and flew in Operation El Dorado Canyon, callsign PUFFY 12. It also flew in Desert Storm. 70-2416 is seen here in July 1987, with the red tail cap of the 494th TFS, 48th TFW. It arrived at AMARC on October 18, 1995. When retired, the aircraft had a total of 5,727.6 flight. (Terry Love Collection)

70-2417 (E2-56) (F-56)
70-2417 was delivered to the USAF on April 27, 1972. It is seen here in July 3, 1989, with the yellow tail cap of the 493rd TFS, 48th TFW. The aircraft was Pave Tack modified and flew in Desert Storm. 70-2417 arrived at AMARC on January 4, 1996. When retired, the aircraft had a total of 5,320.8 flight hours. (Ben Knowles Collection)

70-2418 (E2-57) (F-57)
70-2418 was delivered to the USAF on April 24, 1972. It is seen here in April 21, 1984, with the blue tail cap of the 492nd TFS, 48th TFW. It was Pave Tack modified. While assigned to the 492nd TFS, 48th TFW at RAF Lakenheath, the aircraft crashed and was destroyed on February 23, 1987 near Newmarket, UK. The aircraft experienced a split slab condition (the two sides of the horizontal tail moved in opposite directions) causing loss of control. The air crew successfully ejected, but were injured as a result of failure of the forward parachute bridle holding the parachute to the capsule. This caused the capsule to hit nose first. When the aircraft crashed, it had accumulated 1,425 flights and 3,637.7 flight hours. (Scott Wilson)

70-2419 (E2-58 (F-58)
70-2419 was delivered to the USAF on April 24, 1972. It was Pave Tack modified and flew in Desert Storm. 70-2419 is seen here in August 1994, with the blue tail stripe of the 523rd FS, 27th FW. It arrived at AMARC on April 2, 1996. When retired, the aircraft had a total of 5,482.5 flight hours. (Bob Greby)

71-0883 (E2-59) (F-59)
71-0883 was delivered to the USAF on May 4, 1972. It was Pave Tack modified and flew in Desert Storm. 71-0883 received the Pacer Strike Modification and, as seen here in this photo taken on May 14, 1993, flew with the 524th FS at Cannon AFB. It arrived at AMARC on June 24, 1996. When retired, the aircraft had a total of 5,265.5 flight hours. (Nate Leong)

71-0884 (E2-60) (F-60)
71-0884 was delivered to the USAF on May 9, 1972. It was Pave Tack modified and flew in Desert Storm. 71-0884 received the Pacer Strike Modification and flew with the 524th FS at Cannon AFB. 71-0884 is seen here on May 19, 1995, with the yellow tail stripe of the 524th FS, 27th FW. It arrived at AMARC on May 3, 1996. When retired, the aircraft had a total of 5,017.2 flight hours. (Norris Graser)

CHAPTER THREE: U.S. AIR FORCE TACTICAL F-111s

71-0885 (E2-61) (F-61)
71-0885 was delivered to the USAF on April 20, 1972. It was Pave Tack modified and flew in Desert Storm. 71-0885 is seen here in May 1995, with the blue tail stripe of the 523rd FS, 27th FW. It arrived at AMARC on April 4, 1996. When retired, the aircraft had a total of 6,005.0 flight hours. (Alec Fushi)

71-0886 (E2-62) (F-62)
71-0886 was delivered to the USAF on May 24, 1972. It was Pave Tack modified and flew in Desert Storm. 71-0886 is seen here at McClellan AFB with the green tail cap of the 495th TFS, 48th TFW. It received the Pacer Strike Modification and flew with the 524th FS at Cannon AFB, finally arriving at AMARC on July 22, 1996. When retired, the aircraft had a total of 5,665.8 flight hours. (Author's Collection)

71-0887 (E2-63) (F-63)
71-0887 was delivered to the USAF on May 24, 1972. It was Pave Tack modified and flew in Desert Storm. It is seen here in the markings of the 493rd TFS, 48th TFW. It is carrying an ALQ-131 ECM pod on the forward station and an AN/AQX-14 GBU-15 data link pod on the aft pod station. 71-0887 received the Pacer Strike Modification and flew with the 524th FS at Cannon AFB. It arrived at AMARC on July 8, 1996. When retired, the aircraft had a total of 5,671.8 flight hours. (C. Roger Cripliver Collection)

71-0888 (E2-64) (F-64)
71-0888 was delivered to the USAF on May 18, 1972. It was Pave Tack modified and flew in Operation El Dorado Canyon, callsign LUJAC 22. It also flew in Desert Storm. 71-0888 received the Pacer Strike Modification and flew with the 524th FS at Cannon AFB. It is seen here in May 1996, marked as the 524th FS Flagship. It arrived at AMARC on July 29, 1996. When retired, the aircraft had a total of 5,443.7 flight hours. (Dave Brown)

71-0889 (E2-65) (F-65)
71-0889 was delivered to the USAF on May 31, 1972. It was Pave Tack modified and flew in Operation El Dorado Canyon, callsign KARMA 53. It also flew in Desert Storm. 71-0889 received the Pacer Strike Modification and flew with the 524th FS at Cannon AFB. Seen in this photo taken on September 23, 1995, the aircraft was mismarked as 70-889. It arrived at AMARC on July 8, 1996. When retired, the aircraft had a total of 5,946.3 flight hours. (Keith Snyder)

71-0890 (E2-66) (F-66)
71-0890 was delivered to the USAF on March 26, 1972. It was Pave Tack modified and flew in Desert Storm. 71-0890 is seen here in August 1992, with the yellow tail cap of the 493rd TFS, 48th TFW. received the Pacer Strike Modification and flew with the 524th FS at Cannon AFB. It arrived at AMARC on July 16, 1996. When retired, the aircraft had a total of 5,238.2 flight hours. (Ben Knowles Collection)

CHAPTER THREE: U.S. AIR FORCE TACTICAL F-111s

71-0891 (E2-67) (F-67)
71-0891 was delivered to the USAF on May 31, 1972. It is seen here on September 26, 1972. The aircraft was Pave Tack modified and flew in Desert Storm. 71-0891 received the Pacer Strike Modification and flew with the 524th FS at Cannon AFB. It arrived at AMARC on April 10, 1996. When retired, the aircraft had a total of 5,617.1 flight hours. (USAF)

71-0892 (E2-68) (F-68)
71-0892 was delivered to the USAF on June 28, 1972. It was Pave Tack modified and flew in Desert Storm. 71-0892 arrived at AMARC on April 4, 1996. When retired, the aircraft had a total of 6,132.8 flight hours. (Stephen Miller)

71-0893 (E2-69) (F-69)
71-0893 was delivered to the USAF on June 23, 1972. It was Pave Tack modified and flew in Operation El Dorado Canyon. It attacked Tripoli Airfield and scored hits using high drag MK-82SE on Soviet built transport aircraft. The Pave Tack target video was released to the media. It is seen here on the taxiway at Taif AB, Saudi Arabia. 71-0893 received the Pacer Strike Modification and flew with the 524th FS at Cannon AFB. It arrived at AMARC on July 23, 1996. When the aircraft retired, it had a total of 5,292.3 flight hours. (Terry Love Collection)

71-0894 (E2-70) (F-70)
71-0894 was delivered to the USAF on June 29, 1972. It is seen here on October 7, 1983, while assigned to the 495th TFS, 48th TFW. It has a small white Phantom on the nose, probably painted while visiting an F-4 unit. The aircraft was Pave Tack modified. It arrived at AMARC on November 6, 1995. When the aircraft retired, it had a total of 5,623.7 flight hours. (Scott Wilson) Below: 71-0894 is seen here in May 1995, assigned to the 27th FW at Cannon AFB.(Alec Fushi)

71-0895 to **71-0906** (12 aircraft) were not built after the order was canceled.

72-1441 (E2-71) (F-71)
72-1441 was delivered to the USAF on February 12, 1973. It is seen here in markings of the 390th TFS, 366th TFW. It was the first Pave Tack modified aircraft delivered to the 48th TFW. While assigned to the 494th TFS, 48th TFW, the aircraft crashed and was destroyed on February 4, 1981, while on approach to RAF Lakenheath, UK. The crew ejected successfully without injuries to the crew. The aircraft had a total of 690 flights and 1,808.5 flight hours when it crashed. (Pat Martin Collection)

CHAPTER THREE: U.S. AIR FORCE TACTICAL F-111s

72-1442 (E2-72) (F-72)
72-1442 was delivered to the USAF on March 7, 1973. It was Pave Tack modified and flew in Desert Storm. 72-1442 is seen here in May 1990, with the blue tail cap of the 492nd TFS, 48th TFW. It was later assigned to Eglin AFB. 72-1442 received the Pacer Strike Modification and flew with the 524th FS at Cannon AFB. The aircraft arrived at AMARC on April 10, 1996. When retired, the aircraft had a total of 5,167.5 flight hours. (Craig Kaston Collection)

72-1443 (E2-73) (F-73)
72-1443 was delivered to the USAF on March 28, 1973. It was Pave Tack modified and flew in Desert Storm. 72-1443 is seen here with the red tail stripe of the 494th FS, 48th FW carrying MK 82 AIR 500 pound bombs. It received the Pacer Strike Modification and flew with the 524th FS at Cannon AFB. It arrived at AMARC on July 16, 1996. When retired, the aircraft had a total of 5,692.0 flight hours. (C. Roger Cripliver Collection)

72-1444 (E2-74) (F-74)
72-1444 was delivered to the USAF on April 17, 1973. It was Pave Tack modified and flew in Desert Storm. It's seen here with GBU-10s on the pylons and a Pave Tack in the weapons bay. 72-1444 suffered a major inflight fire on October 29, 1991, while low level. The left engine experienced a catastrophic failure, blowing a large hole in the side of the fuselage. The pilot Captain Craig "Quizmo" Brown was able to zoom the aircraft to 7500 feet, keeping it flying using afterburner on the right engine as necessary. Despite a strong crosswind, the aircraft recovered using a PAR (Precision Radar Approach) to RAF Loffiemouth. It was repaired and returned to flying status. 72-1444 received the Pacer Strike Modification and flew with the 524th FS at Cannon AFB. The aircraft arrived at AMARC on April 10, 1996. When retired, the aircraft had a total of 5,225.9 flight hours. (USAF)

72-1445 (E2-75) (F-75)
72-1445 was delivered to the USAF on May 16, 1973. It was Pave Tack modified and flew in Operation El Dorado Canyon, callsign REMIT 32. It also flew in Desert Storm. It is seen here on April 20, 1994, with the yellow tail cap of the 524th FS, 27th FW. 72-1445 received the Pacer Strike Modification and after the modification was complete, went directly to AMARC from Sacramento ALC. It arrived at AMARC on April 29, 1996. When retired, the aircraft had a total of 4,829.0 flight hours. (Keith Snyder)

72-1446 (E2-76) (F-76)
72-1446 was delivered to the USAF on June 25, 1973. It was Pave Tack modified and flew in Desert Storm. It received the Pacer Strike Modification and flew with the 524th FS at Cannon AFB. 72-1446 is seen here on April 20, 1994, with the yellow tail stripe of the 524th FS, 27th FW. It arrived at AMARC on July 22, 1996. When the aircraft was retired, it had a total of 5,636.8 flight hours. (Douglas Slowiak/Vortex Photo Graphics)

72-1447 (E2-77) (F-77)
72-1447 was delivered to the USAF on July 20, 1973. It is seen here on March 1978. It was Pave Tack modified. While assigned to the 492nd TFS, 48th TFW, the aircraft crashed and was destroyed on June 23, 1982 while flying low level TF near Kinloss, Scotland. The aircraft began violent pitch and roll oscillations, and the crew ejected They received only minor injuries. The aircraft had accumulated 771 flights and 2,016.1 flight hours when it crashed. (K. Kestel)

CHAPTER THREE: U.S. AIR FORCE TACTICAL F-111s

72-1448 (E2-78) (F-78)
72-1448 was delivered to the USAF on August 17, 1973. It is seen here on June 19, 1984, marked as the 48th TFW Flagship. (Scott Wilson) Below: The aircraft was Pave Tack modified and flew in Desert Storm. 72-1448 is seen here in June 1987, marked as the 48th TFW Flagship, with modified markings which the battle streamer added after Operation El Dorado Canyon. It is carrying GBU-24/B 2000 pound low level laser guided bombs. It arrived at AMARC on January 8, 1996. When retired, the aircraft had a total of 5,846.3 flight hours. (Jim Rotramel)

(Adrian Walker)

72-1449 (E2-79) (F-79)
72-1449 was delivered to the USAF on September 24, 1973. It was Pave Tack modified and flew in Operation El Dorado Canyon, callsign LUJAC 21. It also flew in Desert Storm. 72-1449 is seen here in August 1992 marked as the 494th FS Commanders Flagship. It received the Pacer Strike Modification and flew with the 523rd FS at Cannon AFB. It arrived at AMARC on April 10, 1996. When retired, the aircraft had a total of 6,033.9 flight hours. (Tom Kaminski Collection)

72-1450 (E2-80) (F-80)
Above: 72-1450 was delivered to the USAF on October 19, 1973. It was Pave Tack modified and flew in Desert Storm. 72-1450 is seen here on May 11, 1984, with the yellow tail cap of the 493rd TFS, 48th TFW. It received the Pacer Strike Modification and flew with the 524th FS at Cannon AFB. It arrived at AMARC on July 16, 1996. When retired, the aircraft had a total of 5,289.2 flight hours. (Scott Wilson)

72-1451 (E2-81) (F-81)
Right: 72-1451 was delivered to the USAF on November 19, 1973. It is seen here in 1982 with AIM-9 missiles, GBU-10 2000 pound laser guided bombs, a Pave Tack Pod, and an ALQ-131 ECM pod. The aircraft was Pave Tack modified and flew in Desert Storm. 72-1451 arrived at AMARC on January 3, 1996. When retired, the aircraft had a total of 5,379.8 flight hours. (USAF)

72-1452 (E2-82) (F-82)
72-1452 was delivered to the USAF on December 18, 1973. It is seen here on March 7, 1980, with a WA tail code and the yellow and black checkerboard tail stripe of the 57th FWW. The aircraft was Pave Tack modified and flew in Desert Storm. 72-1452 arrived at AMARC on January 4, 1996. When retired, the aircraft had a total flight hours of 5,087.7 flight hours. (Charles B. Mayer)

73-00707 (E2-83) (F-83)
73-00707 was delivered to the USAF on February 26, 1974. It is seen here in September 1986, with the red tail cap of the 494th TFS, 48th TFW. It was Pave Tack modified and flew in Operation El Dorado Canyon, callsign PUFFY 14. The aircraft arrived at AMARC on January 9, 1996. When retired, the aircraft had a total of 5,550.6 flight hours. (Terry Love Collection)

73-00708 (E2-84) (F-84)
73-00708 was delivered to the USAF on April 17, 1974. It is seen here on May 26, 1987, with the green tail cap of the 495th TFS, 48th TFW. The tail cap on this aircraft also has white F-111 silhouettes on the tail cap. The aircraft was Pave Tack modified and flew in Desert Storm. 73-00708 received the Pacer Strike Modification and flew with the 524th FS at Cannon AFB. It arrived at AMARC on July 23, 1996. When retired, the aircraft had a total of 5,438.5 flight hours. (Tom Kaminski Collection)

73-00709 (E2-85) (F-85) NO PHOTO AVAILABLE
73-00709 was delivered to the USAF on June 6, 1974. While assigned to the 366th TFW at Mountain Home AFB, the aircraft crashed and was destroyed on April 21, 1977, at China Lake NAS, California. The aircraft went out of control after a bad split S maneuver. The ejection caused major back injuries to the crew due to the pyrotechnic lines being reversed. This released the forward bridle that holds the parachute to the capsule, causing the capsule to impact the ground nose down. The aircraft had accumulated 272 flights and 738.5 flight hours when it crashed.

73-00710 (E2-86) (F-86)
73-00710 was delivered to the USAF on June 22, 1974. It was Pave Tack modified and flew in Desert Storm. 73-00710 is seen here in September 1977, with the green tail cap of the 495th TFS, 48th TFW. It received the Pacer Strike Modification and flew with the 524th FS at Cannon AFB. It arrived at AMARC on July 16, 1996. When retired, the aircraft had a total of 5,824.4 flight hours. (Terry Love Collection)

73-00711 (E2-87) (F-87)
73-00711 was delivered to the USAF on September 9, 1974. It is seen here in October 1983, with the green tail cap of the 495th TFS, 48th TFW. The aircraft arrived at AMARC on September 5, 1996. When retired, 73-00711 had a total of 5,065.6 flight hours. (Scott Wilson)

73-00712 (E2-88) (F-88)
73-00712 was delivered to the USAF on October 23, 1974. It was Pave Tack modified and flew in Desert Storm. 73-00712 is seen here on April 20, 1994, with the red tail stripe of the 522nd FS, 27th FW. It arrived at AMARC on January 11, 1996. When retired, the aircraft had a total of 5,824.2 flight hours. (Keith Snyder)

73-00713 (E2-89) (F-89)
73-00713 was delivered to the USAF on December 5, 1974. It is seen here with the red tail cap of the 494th TFS, 48th TFW. It was Pave Tack modified and arrived at AMARC on November 13, 1995. When the aircraft retired, it had a total of 5,308.5 flight hours. (Ben Knowles Collection)

CHAPTER THREE: U.S. AIR FORCE TACTICAL F-111s

73-00714 (E2-90) (F-90)
73-00714 was delivered to the USAF on January 31, 1975. It seen here in the markings of the 366th TFW. The aircraft was involved in a mid-air collision with 70-2367 while flying low level over Scotland on April 20, 1979. 73-00714 crashed and was destroyed in Donoch Firth. The crew ejected successfully. The aircraft had accumulated 417 flights and 1,140.0 flight hours when it crashed. (Dave Menard)

73-00715 (E2-91) (F-91)
Right: 73-00715 was delivered to the USAF on March 12, 1975. It was delivered to the USAF on March 12, 1975. 73-00715 crashed on December 12, 1979 at Lakenheath, UK. The aircraft had taken off with a damaged main landing gear. The aircraft came back for a landing with a planned approach end barrier engagement. After touching down, the SOF (Supervisor of Flying) on the ground told the pilot that he had missed the cable, so the pilot added power to go around. However, he had snagged the cable, when the cable had extended to its full length, the aircraft was slammed back onto the runway. The aircraft was sent to AMARC on April 22, 1980. It was rebuilt and put back into service. 73-00715 is seen here on May 7, 1989, with the red tail cap of the 494th TFS, 48th TFW. The aircraft was Pave Tack modified and flew in Desert Storm. 73-00715 arrived at AMARC for the second time on October 18, 1995. When retired, it had a total of 3,880.5 flight hours. (Terry Love Collection)

73-00716 (E2-92) (F-92)
73-00716 was delivered to the USAF on April 15, 1975. It is seen here in September 1981. It was Pave Tack modified. 73-00716 crashed and was destroyed on November 1, 1982, on the Konya Bombing range near Incirlik, Turkey. The loss was caused by an engine fire in the left engine, which burned through the rudder controls, causing loss of control. The crew, 1LT S.J. Bowling and 1LT J.E. Clay, ejected successfully. When the aircraft was lost, it had 729 flights and 1,809.5 flight hours. (Terry Love Collection)

73-00717 (E2-93) (F-93) NO PHOTO AVAILABLE
73-00717 was delivered to the USAF on June 12, 1975. On March 29, 1978, after being hit by lightning, the aircraft crashed and was destroyed at RAF Lakenheath as a result of an abrupt pitch down on landing approach. The aircraft crashed near Mundford, Norfolk, UK. Captain Charles H. Kitchell and 1 Lt Jeffery Moore ejected out of the envelope and were killed. The aircraft had accumulated 285 flights and 785.2 flight hours when it crashed.

73-00718 (E2-94) (F-94) NO PHOTO AVAILABLE
The aircraft was delivered to the USAF on July 23, 1975. 73-00718 crashed and was destroyed on October 5, 1977, near Ramstein AB, Germany during a NATO exercise. As a result of a bombing range weather abort, the aircraft was in a low altitude 180 degree turn when it departed controlled flight (accelerated stall) and crashed. The pilot, Captain Stephen H. Reid and WSO Captain Carl T. Poole, did not attempt to eject and were killed. The aircraft had accumulated 202 flights and 554.5 flight hours when it crashed.

74-0177 (E2-95) (F-95)
74-0177 was delivered to the USAF on September 5, 1975. It is seen here on July 16, 1986, with the blue tail cap of the 492nd TFS, 48th TFW. The aircraft was Pave Tack modified and flew in Operation El Dorado Canyon, callsign JEWEL 61. It also flew in Desert Storm. 74-0177 arrived at AMARC on October 19, 1995. When retired, the aircraft had a total of 5,006.1 flight hours. (B. Morrison)

74-0178 (E2-96) (F-96)
74-0178 was delivered to the USAF on October 10, 1975. It was Pave Tack modified and flew in Operation El Dorado Canyon, callsign REMIT 33. As the 494th TFS Commanders aircraft, it was the lead F-111 in Desert Storm. During Desert Storm, it flew 56 combat sorties in 42 nights. 74-0178 received the Pacer Strike Modification and flew with the 524th FS at Cannon AFB. It is seen here marked as the 27th FW Flagship aircraft. It arrived at AMARC on July 29, 1996, with a total of 5,258.3 flight hours. (Dave Brown)

74-0179 (E2-97) (F-97)
74-0179 was delivered to the USAF on December 4, 1975. It is seen here in August 1979, with the red tail cap of the 494th TFS, 48th TFW. It crashed and was destroyed on September 16, 1982 at RAF Leuchars while on approach for landing. After a Wheel Well Hot fire light and probable fire, hydraulic pressure was lost and while on final approach, the aircraft entered an uncontrolled left roll. The crew ejected safely, receiving only minor injuries. When the aircraft crashed, it had accumulated 714 flights and 1,745.7 flight hours. (Author's Collection)

74-0180 (E2-98) (F-98)
74-0180 was delivered to the USAF on January 15, 1976. It was Pave Tack modified and flew in Desert Storm. 74-0180 received the Pacer Strike Modification and flew with the 523th FS at Cannon AFB, as seen in this photo taken in May 1995. It arrived at AMARC on July 10, 1996. When retired, the aircraft had a total of 4,866.5 flight hours. (Alec Fushi)

74-0181 (E2-99) (F-99)
74-0181 was delivered to the USAF on March 10, 1976. It is seen here in June 1979, with the blue tail cap of the 492nd TFS, 48th TFW. The aircraft was Pave Tack modified and flew in Desert Storm. 74-0181 arrived at AMARC on October 19, 1995. When retired, the aircraft had a total of 4,797.3 flight hours. (Author's Collection)

74-0182 (E2-100) (F-100)
74-0182 was delivered to the USAF on April 20, 1976. It is seen here on May 17, 1980, with the blue tail cap of the 494th TFS, 48th TFW. The aircraft was Pave Tack modified and flew in Desert Storm. 74-0182 arrived at AMARC on April 8, 1996. When retired, the aircraft had a total of 4,433.3 flight hours. (Craig Kaston Collection)

74-0183 (E2-101) (F-101)
74-0183 was delivered to the USAF on June 2, 1976. It is seen here with the green tail cap of the 495th TFS, 48th TFW. The aircraft was the last F-111F modified for Pave Tack. 74-0183 crashed and was destroyed on October 10, 1990, during a Desert Shield training mission over the ASKR Range, north of Taif, Saudi Arabia. The callsign was Cougar 41. Captain Frederick A. Reid and Captain Thomas R. Caldwell were killed. The aircraft had accumulated 1,432 flights and 3,430.1 flight hours when it crashed. (Author's Collection)

74-0184 (E2-102) (F-102)
74-0184 was delivered to the USAF on July 7, 1976. It is seen here in July 1987, with the red tail stripe of the 494th TFS, 48th TFW. The aircraft was Pave Tack modified and flew in Desert Storm. 74-0184 received the Pacer Strike Modification and flew with the 524th FS at Cannon AFB. It arrived at AMARC on July 8, 1996. When retired, the aircraft had a total of 4,949.8 flight hours. (Author's Collection)

CHAPTER THREE: U.S. AIR FORCE TACTICAL F-111s

74-0185 (E2-103) (F-103)
74-0185 was delivered to the USAF on August 18, 1976. It is seen here in August 1988, with the red tail cap of the 494th TFS, 48th TFW. The aircraft was Pave Tack modified and flew in Desert Storm. 74-0185 received the Pacer Strike Modification and flew with the 524th FS at Cannon AFB. It arrived at AMARC on July 10, 1996. When retired, the aircraft had a total of 4586.6 flight hours. (Kevin Foy)

74-0186 (E2-104) (F-104)
74-0186 was delivered to the USAF on September 8, 1976. It was used as the Pave Tack kit proof aircraft. It's seen here during Pave Tack testing. Of interest is the AGM-65 Maverick on the inboard pylon, next to the GBU-15 on the adjacent pylon. While assigned to the 431st Test and Evaluation Squadron, It was used in the testing of the GBU-28 5000 lb class laser guided bomb. The first release of this weapon occurred over the Tonopah Test range in Nevada on February 23, 1991. Four days later it was released by 70-2391 against a target in Iraq. 74-0186 received the Pacer Strike Modification. The aircraft arrived at AMARC on July 10, 1996. When retired, the aircraft had a total of 3,220.3 flight hours. (C. Roger Cripliver Collection)

74-0187 (E2-105) (F-105)
74-0187 was delivered to the USAF on October 19, 1976. It is seen here with the WA tail code and yellow and black checkerboard tail stripe of the 57th FWW. It was Pave Tack modified and received the last Pacer Strike Modification. It was last assigned to the 524th FS at Cannon AFB. The aircraft arrived at AMARC on July 29, 1996. When retired, the aircraft had a total of 2,733.4 flight hours. (Author's Collection)

74-0188 (E2-106) (F-106)
74-0188 was delivered to the USAF on November 22, 1976. It is seen here on May 29, 1982, with the red tail cap of the 494th TFS, 48th TFW. It was Pave Tack modified. 74-0188 crashed and was destroyed on April 26, 1983 in the North Sea off Germany's Borkun Island. The cause of the loss is still unknown. The crew, Captain Charles Vidas and 1 Lt Steven Groak, was killed. The aircraft had accumulated 572 flights and 1,407.1 flight hours when it crashed. (Terry Love Collection)

F-111G

As the B-1B long-range strategic bomber entered service, the FB-111As still remaining with SAC were converted into tactical configuration under the designation F-111G and transferred to Tactical Air Command (TAC).

Under this program the structure was upgraded to withstand 6.5 Gs, G suit connections were installed in the cockpit and weapons release triggers were added to both control sticks. The FB-111A's short range attack missile (SRAM) system for stand-off nuclear delivery was deleted along with removal of SAC's communication equipment including AFSATCOM. The conventional weapons release system was upgraded to provide for nuclear/conventional weapon dual-role capability. Other improvements included the installation of a Have Quick UHF radio and a new ECM system.

The first two F-111G conversions were completed in early 1989. The program was originally scheduled to continue at a rate of approximately twelve conversions per year, but canceled after a total of only 34 F-111Gs were converted from the FB-111As. The first F-111Gs and FB-111As were transferred from SAC to TAC between June and December 1990. Deployment in Europe was considered for a brief time, but the F-111G aircraft were transferred instead to the 27th TFW at Cannon AFB, New Mexico. They were assigned to the 428th Tactical Fighter Training Squadron starting in 1990, augmenting the Wing's F-111Ds in a training role. By mid July 1993, all the F-111Gs had been replaced in the training role by AMP F-111Es returning from Upper Heyford. The F-111G were then surplus to USAF requirements. On June 29, 1993, Australia announced that it was going to purchase 15 of these surplus F-111G aircraft. The aircraft arrived at RAAF Amberley during late 1993 and 1994, joining 18 F-111Cs and four RF-111Cs already in RAAF service. The remainder of the F-111Gs were retired to AMARC during 1993, with the last F-111G, Sacramento ALC's instrumented aircraft (68-0247), arriving at AMARC on December 29, 1993.

F-111G

The 34 F-111Gs started their careers as FB-111As. The FB-111As were upgraded with an avionics update, redesignated as F-111Gs, and transferred from SAC to TAC. The factory numbers listed here are the FB-111A numbers under which they were manufactured and do not represent the order of modification. The factory numbers for the F-111Gs ranged from B1-04 through the last FB-111A built B1-76.

67-0162 (B1-04)
67-0162 is seen here at Dyess AFB in April 1991 marked as the 428th TFTS Flagship and nicknamed *City of Portales*. It arrived at AMARC on June 30, 1993, with a total of 6986.5 flight hours. (Author)

CHAPTER THREE: U.S. AIR FORCE TACTICAL F-111s

67-7193 (B1-07)
67-7193 is seen here at Cannon AFB on May 4, 1991. It arrived at AMARC on March 29, 1993, with a total of 5558.7 flight hours. (Brian C. Rogers)

67-7194 (B1-08)
67-7194 is seen here at Cannon AFB on May 4, 1991. It aircraft arrived at AMARC on March 15, 1993, with a total of 5829.7 flight hours. (Brian C. Rogers)

67-7196 (B1-10) NO PHOTO AVAILABLE
The aircraft arrived at AMARC on March 2, 1993, with a total of 6988.4 flight hours.

68-0241 (B1-13)
68-0241 is seen here at Cannon AFB on May 4, 1991. It arrived at AMARC on June 8, 1993, with a total of 5949.1 flight hours. (Brian C. Rogers)

68-0244 (B1-16)
68-0244 is seen here at Cannon AFB on May 4, 1991. It arrived at AMARC on July 31, 1991, with a total of 6344.5 flight hours. (Brian C. Rogers)

68-0247 (B1-19)
68-0247 is seen here at AMARC on March 12, 1994, with the SM tail code of the 337th Test Squadron SM-ALC. As an instrumented F-111G, 68-0247 was used at Sacramento ALC for engineering flight test work. The aircraft arrived at AMARC on December 29, 1993. When retired, the aircraft had a total of 5480.9 flight hours. (Douglas Slowiak/Vortex Photo Graphics)

68-0252 (B1-24)
The aircraft is seen here at Cannon AFB during April 191. It arrived at AMARC on August 6, 1991, with a total of 5759.2 flight hours. (Gerald McMasters)

68-0254 (B1-26) NO PHOTO AVAILABLE
The aircraft arrived at AMARC on April 29, 1993, with a total of 4649.1 flight hours.

68-0255 (B1-27)
The aircraft arrived at AMARC on September 23, 1993, with a total of 5757.2 flight hours. (Craig Kaston)

68-0257 (B1-29)
68-0257 is seen here at Cannon AFB on May 4, 1991. It arrived at AMARC on April 29, 1993, with a total of 6433.9 flight hours. (Brian C. Rogers)

68-0259 (B1-31)
67-7193 is seen here at Nellis AFB on December 12, 1992. It had a total of 5420.0 USAF flight hours when sold to Australia as A8-259 (Marty Isham)

68-0260 (B1-32)
68-0260 is seen here at Cannon AFB on May 4, 1991. It arrived at AMARC on April 8, 1993, with a total of 6529.8 flight hours. (Brian C. Rogers)

68-0264 (B1-36)
68-0264 is seen here at Cannon AFB on May 4, 1991. It had a total of 6066.6 USAF flight hours when sold to Australia as A8-264. (Brian C. Rogers)

68-0265 (B1-37)
68-0265 is seen here August 17, 1991. The aircraft had a total of 7145.3 USAF flight hours when sold to Australia as A8-265. (Craig Kaston)

68-0270 (B1-42)
68-0270 is seen here at Cannon AFB on May 12, 1992. The aircraft had a total of 5463.2 USAF flight hours when sold to Australia as A8-270. (Craig Kaston)

68-0271 (B1-43)
68-0271 is seen here at Nellis AFB on December 16, 1992. It had a total of 6145.9 USAF flight hours when sold to Australia as A8-271. (Marty Isham)

CHAPTER THREE: U.S. AIR FORCE TACTICAL F-111s

68-0272 (B1-44)
The aircraft had a total of 5766.1 USAF flight hours. The aircraft arrived at AMARC on September 23, 1992. "Boneyard Wrangler" was sold to Australia as A8-272. (Author's Collection)

68-0273 (B1-45)
Seen here at McChord AFB, Washington, 68-0273 arrived at AMARC by truck on May 22, 1991, with a total of 5751.1 flight hours. (Pat Martin)

68-0274 (B1-46)
68-0274 is seen here at Nellis AFB on December 16, 1992. The aircraft had a total of 6056.3 USAF flight hours when sold to Australia as A8-274. (Marty Isham)

68-0276 (B1-48) NO PHOTO AVAILABLE
The aircraft arrived at AMARC by truck on May 22, 1991, with a total of 5334.1 flight hours.

68-0277 (B1-49)
Seen on the flightline at Cannon AFB, 68-0277 had a total of 5980.3 USAF flight hours when sold to Australia as A8-277. The aircraft arrived at Amberley on March 25, 1994. (Craig Kaston)

68-0278 (B1-50)
68-0278 is seen here at Cannon AFB on May 4, 1991. It had a total of 6630.4 USAF flight hours when sold to Australia as A8-278. (Brian C. Rogers)

68-0281 (B1-53)
68-0281 is seen here in 428th FS markings at Cannon AFB on May 12, 1992. The aircraft had a total of 6435.9 USAF flight hours when sold to Australia as A8-281. (Craig Kaston)

68-0282 (B1-54)
68-0282 is seen here at Cannon AFB on May 4, 1991. The aircraft had a total of 7133.8 USAF flight hours when sold to Australia as A8-282. (Brian C. Rogers)

68-0289 (B1-61) NO PHOTO AVAILABLE
The aircraft arrived at AMARC by truck on May 9, 1991, with a total of 5857.5 flight hours.

CHAPTER THREE: U.S. AIR FORCE TACTICAL F-111s

68-0291 (B1-63)
68-0291 is seen here at Cannon AFB on May 12, 1992. The aircraft had a total of 5676.4 USAF flying hours when sold to Australia as A8-291. (Craig Kaston)

69-6503 (B1-65) NO PHOTO AVAILABLE
The aircraft arrived at AMARC by truck on May 22, 1991, with a total of 5653.0 flight hours.

69-6504 (B1-66)
69-6504 is seen here at Cannon AFB on May 4, 1991. The aircraft arrived at AMARC on May 13, 1991, with a total of 6171.8 flight hours. (Brian C. Rogers)

69-6506 (B1-68)
69-6506 is seen here at Fairchild AFB on May 16, 1992. The aircraft had a total of 6105.3 USAF flight hours when sold to Australia as A8-506 (Don Abrahamson)

69-6508 (B1-70)
69-6508 is seen here at Cannon AFB on May 4, 1991. It arrived at AMARC on June 4, 1991, with a total of 6083.0 flight hours. (Brian C. Rogers)

69-6510 (B1-72) NO PHOTO AVAILABLE
The aircraft arrived at AMARC by truck on May 9, 1991, with a total of 6821.0 flight hours.

69-6512 (B1-74)
69-6512 is seen here marked as the 428th FS Flagship on September 29, 1992. The aircraft had a total of 6271.0 USAF flight hours when sold to Australia as A8-512. (Douglas Slowiak/Vortex Photo Graphics)

69-6514 (B1-76)
69-6514 is seen here at Cannon AFB on May 4, 1991. The aircraft had a total of 6093.6 USAF flight hours when sold to Australia as A8-514. (Brian C. Rogers)

CHAPTER FOUR

U.S. Air Force Strategic F-111s

FB-111A

During 1963 through 1965, the slow progress in the Advanced Manned Strategic Aircraft (AMSA) program and fear of earlier-than-expected B-52 failures caused the USAF to look for an interim bomber. One option was to resume B-58 production, which had ended late in 1962, and procure 250 more of these supersonic bombers. B-58 production was considered too costly, so the USAF began considering the F-111A for the role of an interim bomber until the AMSA was fielded. In November 1963, General Dynamics suggested two strategic versions of the F-111. On June 2, 1965, the USAF settled for the least-modified version of the F-111A. This would be the FB-111A interim strategic bomber. The USAF also decided on only 263 FB-111As. Two hundred ten were scheduled to equip 13 squadrons, each with 15 aircraft and a single squadron equipped with 20 aircraft (later reduced to 15) to be used for combat crew training. The remainder of the FB-111As would be used for support and testing. The USAF wanted them quickly, with the first FB-111A operational during fiscal year 1969. SAC wanted more and larger FB-111As, but didn't have the time or the money to spend on the interim FB-111A program. The majority of the future money was to be spent on the larger aircraft, the AMSA, (later developed as the B-1).

Early in 1965, the Office of the Secretary of Defense (OSD) completed a study of the comparative costs and performance of the proposed FB-111A, along with the B-52 and B-58 strategic bombers. This study included the cost effectiveness of a force of 200 FB-111As. Secretary McNamara publicly announced plans to develop the FB-111A on December 10, 1965. This was six months after endorsing the USAF proposal to replace as early as possible 345 B-52s (C through F models) with minimum-modified F-111As.

Reminiscent of Congress' misgivings in November 1962, when General Dynamics, rather than Boeing, was handed the F-111A contract, two factors fueled another round of Congressional concern. The first was replacement of the oldest B-52s by a lesser number of unproven FB-111As. The other was Secretary McNamara's surprise announcement of late 1965, to retire all 80 of the B-58s, SAC's only supersonic bomber, by June 30, 1971. Authorization of the new aircraft was postponed six months, until February 1966, in order to equip the aircraft with more advanced avionics than originally planned. The FB-111A was added to the basic F-111A RDT&E contract (originally issued in May 1964) after Congress had approved a USAF reprogramming request for $26 million of development funds on January 26, 1971.

The purpose of the FB-111A was to provide an interim bomber quickly, with least possible modification, and when available, more advanced avionics could be retrofitted in earlier FB-111A aircraft. The Secretary asked the USAF in January 1966, to begin contract definition on Mark II avionics systems for both the FB-111A and the F-111D, with maximum commonality of the two systems and integration of the planned AGM-69A SRAM missile in the FB-111A's Mark IIB version avionics.

The FB-111A, Weapon System 129A, included extended wing span of 70 feet (a 7-foot increase), stronger landing gear, extra and bigger fuel tanks, and P-7 engines. The P-7 was a new version of the Pratt & Whitney TF30-turbofan engine. It had a maximum thrust of 20,250 pounds with afterburner which was 1,800 pounds more than the P-3 engine of the F-111A and F-111E, but only 100 pounds more than the Navy F-111B's P-12.

The FB-111A also featured the Mark IIB avionic subsystem. This subsystem comprised an improved F-111A attack radar, an inertial navigation system, digital computers, and some advanced displays of the later Mark II that equipped the delayed F-111D. The Mark IIB controlled the new Boeing AGM-69A Short-Range Attack Missile (SRAM).

Basic configuration changes (aimed at increasing the range) were approved and the USAF received extra funding to take care of several other vital SAC needs. Added were weapons bay tanks, turbine starter, horizontal situation display (HSD) and lunar white cockpit lighting. The last two would

FB-111A 67-0162, is seen here on June 14, 1972, during testing at Edwards AFB carrying AGM-69A SRAM (Short Range Attack Missile) CATMs (Captive Air Training Missiles) (Marty Isham Collection)

first enter the 53rd FB-111A on the production line which was programmed as the initial aircraft of the second operational wing. Retrofit of earlier FB-111A productions was not planned, but SAC intended to request a retrofit modification later.

One of the major problems of the future FB-111A was the engines. The TF30-P3 engines of the tactical F-111As (and the F-111Es), with 18,500 pounds of thrust each, were inadequate for the heavier FB-111A. The Navy F-111B's new P-12 engine, with 20,000 pounds of thrust, appeared more promising, but it was just entering service in November 1966, and would not be available in time. By mid-1967, the USAF had selected the P-12 as the engine for the FB-111A. It was to have been configured with semi-actuator ejector (SAE) nozzles and known as the P-5. Development of the P-12/P-5 engine was tied to U.S. Navy procurement of the F-111B. Pratt and Whitney, however, lacked a firm production go-ahead for the P-12 due to consideration of elimination of the Navy F-111B. Instead of the P-12, a P-7 engine, with 20,250 pounds of thrust, was be developed for the FB-111A.

A modified RDT&E F-111A (F-111 aircraft No. 18, 63-9783), still equipped with TF30-P-1 engines and the tactical F-111A landing gear, served as FB-111A prototype. The aircraft flew for 45 minutes on its maiden flight on July 30, 1967. It achieved Mach 2 on its first flight. Accepted by the USAF after the first flight, it remained with General Dynamics for more testing. Category I testing, a prime contractor's responsibility, started on July 19, 1967, and lasted through November 1971.

The range was still not what SAC had hoped. The FB-111A's comparatively short range for a strategic aircraft was inherent in aircraft which had been converted from tactical designs. Additionally, the Triple Plow II air diverter, designed to prevent engine stall, decreased its range even more. To extend the combat range, larger fuel tanks, external tanks, and air refueling would have to be used. Other FB-111A changes, including redesign of the aft fuselage, limited the FB-111A's maximum speed to around Mach 2.

The first production FB-111A (67-0159) flew on July 13, 1968 and was accepted by the USAF on August 30. A second FB-111A production aircraft (67-0160) was accepted on October 25. Subsystem problems, mainly with the Mark IIB, slowed further deliveries. The initial Mark II avionic units had been delivered to General Dynamics on November 21, 1967. Flight Tests of the Mark IIB system, starting on March 31, 1968 using a modified F-111A (F-111 No. 25, 65-5707), showed good results. Problems surfaced during the first full system test in June when various components began to interfere with each other. SRAM system tests were accomplished on a modified F-111A, 66-0011.

The USAF accepted its next FB-111A on June 23, 1969. This third FB-111A (67-0161) differed from the previous two in that it featured a fully developed Triple Plow II air diverter, a complete Mark IIB avionics system, and the new P-7 engines.

Because of the sophistication of the FB-111As systems, the USAF raised the number of aircraft to be used in the formal testing program to seven, using additional modified F-111As (63-9783 and 65-5707) for a few special tests. These seven test aircraft included first six FB-111A production models which would revert to their operational combat aircraft following testing. Category II testing started on September 4, 1968, (14 months after the beginning of the Category I tests) at Edwards AFB. Category II tests were still going on when Category III testing started in October 1971, and had not been completed when Category III testing ended on July 31, 1972. The third FB-111A production aircraft was also allocated to the Category II tests. The Category III tests were conducted at Pease AFB, New Hampshire. The cold weather in New Hampshire highlighted some problems not evident during testing in Edwards desert environment. Brakes failed to work in the cold, as the brake fluid froze. Because of poor insulation, frozen valves prevented transfer of fuel from auxiliary to main tanks. These problems were quickly fixed.

A program of 263 planes was projected when the FB-111A development began. This dropped to 126 on November 28, 1968, because of problems with the basic F-111, production delays, and rising costs. The reduction followed cancellation of the F-111K and the Navy F-111B. The cost of 263 FB-111As in 1966, was estimated at $81.7 billion. The second and final cut took place in March 1969, when the total FB-111A purchase dipped to 76. With the decrease in aircraft from 263 to 76 FB-111As, the unit cost per aircraft soared from $6.45 million to $12.93 million.

The 7th FB-111A (67-7193) entered operational service on October 8, 1969. This aircraft was SAC's first new strategic bomber design since the initial B-58 was accepted on August 1, 1960. This FB-111A had actually been assigned to the 340th Bomb Group since September 25,1969. The next 14 aircraft were assigned to the 9th Bombardment Squadron Medium (BS) of the 340th Bomb Group at Carswell. The 4007th CCTS was responsible for initial FB-111A combat crew academic training, with the 9th BS conducting the flying training.

Problems with the FB-111A's wing longerons and terrain-following radar slowed production. The 4007th CCTS was still short seven aircraft when all deliveries from General Dynamics were stopped. The few FB-111As already delivered were returned to General Dynamics for rework due to the mandatory Recovery Program caused by the F-111A crash of December 1969. In April 1970, the first of CCTS FB-111A left Carswell to undergo a 75-day test and structural inspection and receive necessary modifications. It was ready for assignment back to the 4007th in July 1970. In September 1970 the 4007th CCTS received the last of its 15 FB-111As after the final slippage and resulting Recovery Program modifications.

The FB-111A testing of the SRAM began on March 27, 1970. The first separation of a dummy air-to-surface SRAM

F-111A A-29 (66-0011) was used for SRAM System testing. It's seen here carrying four SRAM test missiles. The fairings for the launch recording cameras are visible on the wing tips, on the nose just in front of the nose gear doors, and mid fuselage just aft of the main gear doors. (John Cook Collection)

missile from an FB-111A (at Mach 0.9 and 25,000 feet altitude) occurred on October 19, 1968, at Eglin AFB. First launch of an operational SRAM from an FB-111A occurred in 1974. The FB-111A/SRAM integration test program had 15 successful launches out of 19, well worth the $140 million spent for the FB-111A integration program.

The 509th Bombardment Wing, Heavy was redesignated the 509th Bombardment Wing, Medium, on December 1, 1969 in preparation for arrival of FB-111As. On December 16, 1970, the 509th Bomb Wing at Pease AFB, received its first FB-111A. The 509th, after many difficulties, was fully combat ready in October 1971. The 380th at Plattsburgh AFB, New York,

Two 509th Bomb Wing FB-111As (69-6508 and 69-6510) are seen here over Portsmouth Bay, New Hampshire, in September 1988. (Chris McWilliams Collection)

converted to FB-111A medium bombers in 1971, and began flying FB-111A strategic bombardment missions in July 1971. The training mission, along with the 4007th CCTS, was relocated from Carswell to Plattsburgh and became part of the 380th Strategic Aerospace Wing on December 31, 1971. FB-111A combat crew training was commenced FB-111A in August 1971. The 4007th CCTS retained its original designation until redesignated as the 530th CCTS on July 1, 1986. The 380th Strategic Aerospace Wing (the second of SAC's only two wings of FB-111As) at Plattsburgh AFB, New York, became combat ready during 1972. The 380th won the SAC bombing and navigation competition, and Fairchild Trophy, in 1974, 1976, and 1977.

To increase survivability, FB-111As of both Bomb Wings stood satellite alert during the early 1970s. The 380th Bomb Wing sent some of their jets on a programmed 90 day rotation to Kincheloe AFB, Michigan and K.I. Sawyer AFB, Michigan. The 509th Bomb Wing sent some of their jets to Rickenbacker AFB, Ohio and McGuire AFB, New Jersey. The actual deployment time for the aircraft turned out to be closer to 70 days. Satellite alert was discontinued in April 1975.

FB-111A production ended on June 1, 1971. The USAF accepted 69-6514, the last FB-111A, on June 30, 1971. When FB-111A deliveries ended, a total of 76 FB-111As had been built, with 75 seeing operational service (the 55th FB-111A aircraft, 68-0283 crashed before delivery on October 7, 1970). A 77th aircraft was an F-111A, (63-9783) modified and charged to the FB-111A program. 63-9783 was accepted in fiscal year (FY) 68. The USAF accepted three FB-111As in FY 69 (two in the fall of 1968 and one in June 1969); six in FY 70 (between July and December 1969, when all F-111s were grounded); and 66 in FY 71 (between August 1970, when the grounding was lifted, and June 1971).

The production cost of $9.8 million per aircraft was broken down as follows: airframe – $4,201,000; engines (installed) – $1,735,000; electronics – $2,550,000 and armament – $1,342,000. This cost did not include $2,043,000 of RDT&E costs and $628,811 worth of medication per bomber. In mid-1973 the actual cost of each FB-111A was set at $12.5 million – $400,000 less than anticipated in late 1969.

A number of factors – the end of the cold war, the SALT Treaty, the B-1B and B-2, and most of all a decreasing military budget, spelled the end of the FB-111A. The 509th Bomb Wing began giving up FB-111As (many being converted to F-111G configuration) in June 1990, with the last FB-111As leaving on September 30, 1990. The Wing's two Bomb Squadrons, the 393rd and the 715th were inactivated on September 30, 1990. The 393rd BS was activated three years later, as part of the "new" 509th Bomb Wing at Whiteman AFB, as the first B-2 squadron. The 380th Bomb Wing began giving up its FB-111As in mid-1991 with the 528th BS being inactivated on July 1, 1991, and the 529th inactivated on September 30, 1991.

Thirty-four of the FB-111As were modified to F-111Gs configuration and assigned to the 27th TFW at Cannon AFB. The remaining FB-111As were retired to AMARC during the summer of 1991. Nine FB-111As were put on display at Air Force bases in the U.S.

509th BW FB-111A 67-7196 is seen here over Portsmouth Bay, New Hampshire in October 1986, carrying two AGM-69A SRAM CATMs. (Don Sutherland)

CHAPTER FOUR: U.S. AIR FORCE STRATEGIC F-111s

FB-111A

Seventy-Six FB-111As were built, with F-111As 63-9783, 65-5707, and 66-0011 used in the FB-111A test program. The factory numbers for the FB-111As ranged from B1-01 through B1-76. The first operational FB-111As were delivered to the 4007th CCTS of the 340th Bomb Group at Carswell AFB, Fort Worth, Texas. The operational FB-111As were assigned to the 380th Bombardment Wing (Medium) at Plattsburgh AFB, New York, and the 509th Bombardment Wing (Medium) at Pease AFB, New Hampshire. Thirty-four of the FB-111As were modified to F-111Gs, with the remainder of the FB-111As retiring to AMARC or being posted as gate guards or museum aircraft.

67-0159 (B1-01)
67-0159, the first FB-111A, was delivered to the USAF on August 30, 1968. It was retired to McClellan AFB, California Museum. The aircraft was manufactured with Triple Plow I intakes instead of Triple Plow II intakes used on the remainder of the FB-111As. As seen here in August 1985, while assigned to AFFTC 67-0159 was painted in a bright orange/white scheme during later flight tests. Following initial flight tests at Edwards AFB, 67-0159 spent most of its active career at Sacramento ALC. When retired on November 15, 1991, it had accumulated 927 flights and 2,038.8 flight hours. (Ben Knowles Collection)

67-0160 (B1-02)
Above: 67-0160 was delivered to the USAF on October 25, 1968. It is seen here at Edwards AFB on September 18, 1974. The aircraft was fitted with preliminary design of double blow-in doors on the intake cowls. This inlet system was called the Super Plow inlet system and featured double blow-in doors. It was assigned to AFFTC for flight testing. 67-0160 was decommissioned on February 11, 1974, and had accumulated 349 flights and 595.8 flight hours. The aft section was removed and was used in 1978 to repair 67-7194. The remainder of the aircraft was shipped to AMARC, arriving on February 12, 1990. (Tom Brewer)

67-0161 (B1-03)
Right: 67-0161 was delivered to the USAF on June 23, 1969. This was the first FB-111A to have standard Triple Plow II intakes, a complete Mark IIB avionics system, and the new P-7 engines. It was transferred to the 380th Bomb Wing in August 1978. While assigned to the 380th Bomb Wing, it carried the nickname and nose art *Liquidator*. The aircraft arrived at AMARC on May 6, 1991. When retired, the aircraft had a total of 5,419.8 flight hours. (Terry Love Collection)

67-0162 (B1-04)
67-0162 was delivered to the USAF on September 18, 1970. It was used for Category II testing and later assigned to the 380th Bomb Wing at Plattsburgh AFB. During Bomb Competition, 67-0162, as seen in this photo taken at Barksdale AFB on November 14, 1974, carried the *Apple 1* insignia for Giant Voice 1974. 67-0162 was later named *Nocturnal Mission*. It was converted to an F-111G. (Ray Leader)

67-0163 (B1-05)
67-0163 was delivered to the USAF on July 24, 1969. It was initially used in Category II testing as the SRAM test aircraft. It's seen here on a test flight carrying four external SRAMs. The AGM-69A SRAM, with its nuclear warhead, range up to 100 miles, very small radar cross-section and multiple Mach speed, once launched was impossible to defend against. While assigned to the 380th Bomb Wing at Plattsburgh AFB, the aircraft carried the nickname and nose art *Moonlight Maid*. The aircraft arrived at AMARC on July 2, 1991. When retired, the aircraft had a total of 6,480.4 flight hours (USAF)

67-7192 (B1-06)
67-7192 was delivered to the USAF on November 21, 1969. It was initially used in Category II testing before being assigned to the 380th Bomb Wing. In this photo it is carrying cluster bombs (CBUs). While assigned to the 380th Bomb Wing at Plattsburgh AFB, 67-7192 carried the nickname and nose art *Slightly Dangerous*. The aircraft arrived at AMARC on July 2, 1991. When retired, 67-7192 had a total of 6,360.1 flight hours. (USAF Via Tom Brewer)

CHAPTER FOUR: U.S. AIR FORCE STRATEGIC F-111s

67-7193 (B1-07)
67-7193 was delivered to the USAF on September 28, 1969. It was the first FB-111A delivered to SAC's 4007th CCTS, 340th Bomb Group at Carswell AFB, Texas on September 29, 1969. The aircraft was assigned to the 509th Bomb Wing, Pease AFB, New Hampshire. It's seen here on July, 29, 1984. 67-7193 carried the nickname and nose gear door art *Tiger Lil*. The aircraft was converted to an F-111G. (Brian C. Rogers)

67-7194 (B1-08)
67-7194 was delivered to the USAF on November 4, 1969. It is seen here in May 1984. While assigned to the 509th Bomb Wing, 67-7194 ran off the runway during a night landing. The aircraft was rebuilt and returned to flying status. Once again 67-7194 sustained major damage, this time from hitting the approach lights and the resulting hard landing at Pease AFB on February 25, 1976. The aircraft was rebuilt at General Dynamics Restoration Facility, Fort Worth using parts of other Aardvarks. The rebuild started on September 1, 1978, with front section from 67-7194 being mated to the aft section of 67-0160, along with a spare vertical stabilizer from FB-111A 69-6513. Because of the use of other F-111 parts in the rebuild, the 67-7194 was nicknamed Frankenvark. The aircraft returned to flying status assigned to the 380th Bomb Wing in September 1980. The aircraft was converted to an F-111G. (Doug Remington)

67-7195 (B1-09)
67-7195 was delivered to the USAF on November 30, 1969. While assigned to the 509th Bomb Wing at Pease AFB, the aircraft carried the nicknames *Dave's Dream* and *Big Stink*. It's seen here at Ellsworth AFB on July 17, 1989, with the nose gear door art of *Big Stink*. The aircraft arrived at AMARC on May 28, 1991. When retired, the aircraft had a total of 5,483.5 flight hours. (Brian C. Rogers)

67-7196 (B1-10)
67-7196 was delivered to the USAF on August 30, 1970. It is seen here in October 1986, over Portsmouth New Hampshire carrying a practice SRAM under each wing. The aircraft was assigned to the 509th Bomb Wing, Pease AFB, New Hampshire. 67-7196 carried the nickname and nose gear door art *Ruptured Duck* from Captain Ted Lawson's B-25 Tokyo raider. The aircraft was converted to an F-111G. (Don Sutherland)

68-0239 (B1-11)
68-0239 was delivered to the USAF on December 9, 1969. While assigned to the 380th Bomb Wing at Plattsburgh AFB, the aircraft carried the nickname and nose art *Rough Night*. 68-0239 is seen here on July 2, 1991. It is preserved on display at K.I. Sawyer AFB, Michigan, one of the satellite alert facilities of the 380th Bomb Wing. On July 11, 1991, the aircraft was flown to KI Sawyer for static display. This aircraft flew 6,122 hours in 22 years of service. (Andrew H. Cline)

68-0240 (B1-12)
68-0240 was delivered to the USAF on August 24, 1970. Seen here in May 1983, 68-0240 was assigned to the 380th Bomb Wing at Plattsburgh AFB. It arrived at AMARC on July 3, 1991. When retired, the aircraft had a total of 6041.2 flight hours. (Terry Love Collection)

CHAPTER FOUR: U.S. AIR FORCE STRATEGIC F-111s

68-0241 (B1-13)
68-0241 was delivered to the USAF on September 30, 1970. While assigned to the 380th Bomb Wing at Plattsburgh AFB, 68-0241, as seen here in June 1989, carried the nickname and nose art *Undecided*. The aircraft was converted to an F-111G. (Craig Kaston Collection)

68-0242 (B1-14)
The aircraft was delivered to the USAF on September 22, 1970. 68-0242 was assigned to the 509th Bomb Wing, Pease AFB, New Hampshire. It's seen here in March 1981, carrying MK 82 500 pound bombs. During a Red Flag mission on June 7, 1983, the aircraft crashed and was destroyed near the Nevada – Arizona border. The aircrew heard a loud thump, followed by a left engine fire warning light. The fire could not be extinguished. The crew ejected after losing aircraft control following burn through of the rudder controls. When the aircraft crashed, it had accumulated 978 flights and 3,564.6 flight hours. (Terry Love Collection)

68-0243 (B1-15)
Right: The aircraft was delivered to the USAF on August 31, 1970. 68-0243 is seen here at McCoy AFB, Orlando, Florida in November 1971. It's marked with the winged 2 of the 2nd Air Force. While assigned to the 380th Bomb Wing at Plattsburgh AFB, 68-0243 carried the nickname and nose art *Jungle Queen*. The aircraft crashed and was destroyed on February 2, 1989, near St Johnsburg, Vermont. The aircraft lost control as the result of an external fuel tank on the right inboard pivot pylon swinging outward shortly after take-off. The crew ejected successfully. The aircraft had accumulated 1,302 flights and 4,459.2 flight hours at the time on the crash. (Ken Buchanan via Tom Brewer)

68-0244 (B1-16)
68-0244 was delivered to the USAF on September 28, 1970. While assigned to the 380th Bomb Wing at Plattsburgh AFB, 68-0244 is seen here at Carswell AFB on May 24, 1986. It carried the nickname and nose art *Lucky Strike* on the right side and *The Flying Circus/380th Bomb Group Association* art on the left fuselage side. The aircraft was converted to an F-111G. (Brian C. Rogers)

68-0245 (B1-17)
68-0245 was delivered to the USAF on September 21, 1970. It is seen here at Carswell AFB on November 4, 1984. While assigned to the 380th Bomb Wing at Plattsburgh AFB, the aircraft carried the nickname and nose art *Ready Teddy*. 68-0245 is now on display at March AFB, California. (Brian C. Rogers)

68-0246 (B1-18)
68-0246 was delivered to the USAF on September 18, 1970. It is seen here at Offutt AFB in March 1984. While assigned to the 380th Bomb Wing, 68-0246 carried the nickname and nose art *Royal Flush*. While assigned to the 509th Bomb Wing, it was nicknamed *Pistol Packin Mama*. The aircraft arrived at AMARC on July 1, 1991. When retired, the aircraft had a total of 6,304.7 flight hours. (George Cockle)

CHAPTER FOUR: U.S. AIR FORCE STRATEGIC F-111s

68-0247 (B1-19)
68-0247 was delivered to the USAF on September 30, 1970. It is seen here in June 1978, with its tail painted for the Tiger Meet. 68-0247 was assigned to the 393rd Bomb Squadron 509th Bomb Wing. While assigned to the 380th Bomb Wing at Plattsburgh AFB, the aircraft carried the nickname and nose art *Missouri Miss,* and *Fort Worth Gal* when assigned to the 509th Bomb Wing. (Ted Van Geffen)

68-0247 is seen here on August 23, 1989, in Sacramento Air Material Area (SMAMA) engineering flight test markings. It was one of two FB-111As used as testbeds for the FB-111A AMP program. The aircraft was converted to an F-111G and remained at Sacramento until retired. (Scott Wilson)

68-0248 (B1-20)
68-0248 was delivered to the USAF on October 10, 1970. While assigned to the 380th Bomb Wing at Plattsburgh AFB, 68-0248 carried the nickname and nose art *Free For All*. The aircraft is seen here at Ellsworth in July 1995, being prepared to go on display at the South Dakota Air and Space Museum located at Ellsworth AFB, South Dakota. (Author)

68-0249 (B1-21)
68-0249 was delivered to the USAF on September 20, 1970. It is seen here in February 1983. While assigned to the 380th Bomb Wing, it carried the nickname and nose art *Little Joe*. 68-0249 was the last operational FB-111A to leave Plattsburgh, arriving at AMARC on July 10, 1991. When retired, the aircraft had a total of 6,104.5 flight hours. (Ben Knowles Collection)

68-0250 (B1-22)
68-0250 was delivered to the USAF on October 9, 1970. It is seen here at Offutt AFB in May 1984. While assigned to the 380th Bomb Wing at Plattsburgh AFB, the aircraft carried the nickname and nose art *Silver Lady*. 68-0250 arrived at AMARC on July 3, 1991. When retired, the aircraft had a total of 5,718.0 flight hours. (George Cockle)

68-0251 (B1-23)
68-0251 was delivered to the USAF on March 26, 1971. It is seen here at Carswell AFB on April 1, 1988. While assigned to the 380th Bomb Wing at Plattsburgh AFB, 68-0251 carried the nickname and nose art *Shy-Chi Baby*. The aircraft arrived at AMARC on July 1, 1991. When retired, the aircraft had a total of 5539.8 flight hours. (Brian C. Rogers)

CHAPTER FOUR: U.S. AIR FORCE STRATEGIC F-111s

68-0252 (B1-24)
68-0252 was delivered to the USAF on October 8, 1970. It is seen here on June 12, 1976. While assigned to the 380th Bomb Wing at Plattsburgh AFB, the aircraft carried the nickname and nose art *Six Bits*. 68-0252 was converted to an F-111G. (Ray Leader)

68-0253 (B1-25) NO PHOTO AVAILABLE
68-0253 was delivered to the USAF on August 22, 1970. 68-0253 crashed and was destroyed on October 7, 1970, near Carswell AFB, while assigned to the 340th Bomb Group. The aircraft lost control and crashed when it entered the Ground Controlled Approach (GCA) pattern without its flaps and slats extended. Lt Col Robert Montgomery and Lt Col Charles Robinson were killed when ejection was attempted out of the envelope. The aircraft had logged 26 flights and 106.0 flight hours when it crashed.

68-0254 (B1-26)
Right: 68-0254 was delivered to the USAF on August 28, 1970. It was assigned to the 380th Bomb Wing at Plattsburgh AFB, and nicknamed *Pappy's Passion*. As seen here, 68-0254 was assigned to AFFTC at Edwards AFB. During 1990 and 1991 it was used as part of F-111 Digital Flight Control System testing. The aircraft was converted to an F-111G. (USAF via Marty Isham)

68-0255 (B1-27)
68-0255 was delivered to the USAF on October 21, 1970. While assigned to the 380th Bomb Wing, as seen here at Ellsworth AFB on August 10, 1989, it carried the nickname and nose art *Sleepy Time Gal*. While 68-0255 was assigned to the 509th Bomb Wing, it carried the nickname *Untouchable*. The aircraft was converted to an F-111G. (Brian C. Rogers)

68-0256 (B1-28)
68-0256 was delivered to the USAF on August 28, 1970. It is seen here at Sacramento-ALC in March 1980. While assigned to the 380th Bomb Wing at Plattsburgh AFB, 68-0265 was nicknamed *The Screamer*. 68-0256 arrived at AMARC on July 9, 1991. When retired, the aircraft had a total of 6,002.9 flight hours. (Charles B. Mayer)

68-0257 (B1-29)
68-0257 was delivered to the USAF on September 13, 1970. It is seen here at Peterson Field, Colorado in May 1985. While assigned to the 380th Bomb Wing, it carried the nickname and nose art *Maid In The USA*. While assigned to the 509th, 68-0257 carried the nickname *Next Objective*. The aircraft was converted to an F-111G. (Chuck Robbins)

68-0258 (B1-30)
68-0258 was delivered to the USAF on September 12, 1970. It is seen here on March 26, 1991. While assigned to the 380th Bomb Wing, it carried the nickname *Hell's Belles*. While assigned to the 509th Bomb Wing, 68-0258 was nicknamed *Wild Hare*. It arrived at AMARC on July 3, 1991. When retired, the aircraft had a total of 6,263.8 flight hours. (Craig Kaston)

CHAPTER FOUR: U.S. AIR FORCE STRATEGIC F-111s

68-0259 (B1-31)
68-0259 was delivered to the USAF on August 30, 1970. It is seen here on September 7, 1981. The aircraft was assigned to the 509th Bomb Wing, Pease AFB, New Hampshire. 68-0259 was nicknamed *Necessary Evil* while assigned to the 509th BW. It was converted to an F-111G and later sold to Australia as A8-259. (Brian C. Rogers)

68-0260 (B1-32)
68-0260 was delivered to the USAF on September 11, 1970. It is seen here on June 9, 1984. It was assigned to the 380th Bomb Wing at Plattsburgh AFB and was nicknamed *SNAFU*. 68-0260 was converted to F-111G. (Brian C. Rogers)

68-0261 (B1-33)
68-0261 is seen here on June 25, 1979. The aircraft was delivered to the USAF on September 27, 1970. 68-0261 was assigned to the 509th Bomb Wing, Pease AFB, New Hampshire. The aircraft crashed and was destroyed on September 18, 1979, while on a Red Flag mission over the Nellis Ranges. Captains Phillip Donovan and William Full did not attempt ejection and were killed. The aircraft had flown 684 flights and accumulated 2,560.0 flight hours. (Author's Collection)

68-0262 (B1-34)
68-0262 was delivered to the USAF on October 11, 1970. It is seen here at Offutt AFB in September 1982. While assigned to the 380th Bomb Wing at Plattsburgh AFB, the aircraft carried the nickname and nose art *Lady Luck*. 68-0262 arrived at AMARC July 9, 1991. When retired, the aircraft had a total of 6,221.6 flight hours. (George Cockle)

68-0263 (B1-35)
68-0263 was delivered to the USAF on October 30, 1970. Seen here on landing approach, it was assigned to the 509th Bomb Wing, Pease AFB, New Hampshire. 68-0273 crashed and was destroyed on January 30, 1981, at Portsmouth, New Hampshire. While on a Functional Check Flight, the aircraft experienced an uncommanded roll and crashed. The crew ejected successfully. The aircraft had flown 682 flights and logged 2,472.7 flight hours when it crashed. (Author's Collection)

68-0264 (B1-36)
68-0264 was delivered to the USAF on October 19, 1970. It is seen here on June 7, 1975. While assigned to the 380th Bomb Wing at Plattsburgh AFB, the aircraft carried the nickname and nose art *Jesebelle*. This art was named after 380th Bomb Group WWII B-24 (42-72953). 68-0264 was converted to an F-111G and sold to Australia as A8-264. (James M. Lukas)

68-0265 (B1-37)
68-0265 was delivered to the USAF on October 28, 1970. It is seen here on June 6, 1987. While assigned to the 380th Bomb Wing at Plattsburgh AFB, 68-0265 carried the nickname and nose art *Angel in de Skies*, and also *Net Results*. The aircraft was converted to an F-111G and sold to Australia as A8-265. (F. A. Jackson)

CHAPTER FOUR: U.S. AIR FORCE STRATEGIC F-111s

68-0266 (B1-38)
68-0266 was delivered to the USAF on October 29, 1970. It is seen here at McCoy AFB, Orlando, Florida on December 14, 1971. It's marked with the winged 2 of the 2nd Air Force. It was assigned to the 509th Bomb Wing, Pease AFB, New Hampshire. It crashed and was destroyed on February 14, 1977. Following an inflight refueling, the aircraft descended one turn too soon and hit a mountain near Bristol Tennessee, killing the crew, Captain Edward R. Riley and Captain Jeremiah E. Sheehan. The aircraft had accumulated 436 flights and 1,650.2 flight hours. (Bill Strandberg)

68-0267 (B1-39)
68-0267 was delivered to the USAF on November 15, 1970. It is seen here at Ellsworth AFB on May 23, 1991. The aircraft was assigned to the 509th Bomb Wing, Pease AFB, New Hampshire and was nicknamed *Memphis Belle II*. While assigned to the 380th Bomb Wing, 68-0267 was nicknamed *Pom Pom Express*. It is now on display at the SAC Museum, Offutt AFB, Omaha, Nebraska. (Brian C. Rogers)

68-0268 (B1-40)
Right: 68-0268 was delivered to the USAF on November 17, 1970. 68-0268 is seen here on November 31, 1979. The aircraft was assigned to the 509th Bomb Wing, when it crashed into the ocean (at a depth of 180 feet) and was destroyed on October 6, 1980, near Bangor, Maine. The cause of the loss could not be determined. Major Thomas Mullen and Captain Gary Davis were killed. The aircraft had accumulated 680 flights and logged 2,446.0 flight hours when it crashed. (Ben Knowles)

68-0269 (B1-41)
68-0269 was delivered to the USAF on November 14, 1970. It is seen here on August 11, 1979. It was assigned to the 509th Bomb Wing and named *New Hampshire Special* for the 1974 Giant Voice Bomb Competition. While assigned to the 380th Bomb Wing at Plattsburgh AFB, 68-0269 carried the nickname and nose-art *Sad Sack*. The aircraft arrived at AMARC on July 9, 1991. When retired, the aircraft had a total of 5,648.9 flight hours. (Craig Kaston Collection)

68-0270 (B1-42)
68-0270 was delivered to the USAF on November 25, 1970. It is seen here at Carswell AFB on January 13, 1986. While assigned to the 380th Bomb Wing at Plattsburgh AFB, 68-0270 carried the nickname and nose art *Full House*. The art was the same as a 509th Composite Group WWII B-29 (44-27298) which flew as the weather plane on the Hiroshima bomb mission. The aircraft was converted to an F-111G and sold to Australia as A8-270. (Brian C. Rogers)

68-0271 (B1-43)
68-0271 was delivered to the USAF on November 15, 1970. It is seen here in June 1983. While assigned to the 380th Bomb Wing, 68-0271 carried the nickname *On De-Fence*. The name *Boomerang* was given to 68-0271 while assigned to the 509th BW. The aircraft was converted to an F-111G and sold to Australia as A8-271. (Terry Love Collection)

68-0272 (B1-44)
68-0272 was delivered to the USAF on November 17, 1970. It is seen here at Pease AFB, New Hampshire with the tiger stripe painted rudder carried on some aircraft of the 393rd Bomb Squadron Tigers. The aircraft carried the nickname and nose gear door art *Wild Hare* while assigned to the 509th Bomb Wing, Pease AFB. It was used as an AMP testbed at Sacramento ALC during 1986 and 1987. 68-0272 was converted to an F-111G and sold to Australia as A8-272. (F. A. Jackson)

68-0273 (B1-45)
68-0273 was delivered to the USAF on November 23, 1970. It is seen here at Elmendorf AFB, Alaska in July 1989. The aircraft was assigned to the 509th Bomb Wing, Pease AFB, New Hampshire, and carried the nickname and nose gear door art *Milk Wagon*. 68-0273 was converted to an F-111G. (Jim Goodall)

68-0274 (B1-46)
68-0274 was delivered to the USAF on December 8, 1970. It is seen here on June 27, 1986. While assigned to the 509th Bomb Wing, it carried the nickname *Laggin' Dragon*. After being assigned to the 380th Bomb Wing, 68-0274 was assigned to Det 3, 431 TES at McClellan AFB. The aircraft was converted to an F-111G and sold to Australia as A8-274. (Brian C. Rogers)

68-0275 (B1-47)
68-0275 was delivered to the USAF on November 30, 1970. It is seen here on May 3, 1987. While assigned to the 509th Bomb Wing, 68-0275 carried the nickname and nose gear door art *Bomble Bee*. It is now on display at Kelly AFB, Texas. (Douglas Slowiak/Vortex Photo Graphics)

68-0276 (B1-48)
68-0276 was delivered to the USAF on December 24, 1970. It is seen here at Offutt AFB. The aircraft was assigned to the 509th Bomb Wing, Pease AFB, New Hampshire. It carried the nickname and nose gear door art *Gruesome Goose*. The aircraft was converted to an F-111G. (Don McGarry)

68-0277 (B1-49)
68-0277 was delivered to the USAF on December 24, 1970. It is seen here at Forbes Field ANGB in March 1987. While assigned to the 380th Bomb Wing at Plattsburgh AFB, the aircraft carried the nickname and nose art *Double Trouble* which was named after 380th Bomb Group WWII B-24. The aircraft was converted to an F-111G and sold to Australia as A8-277. (Jerry Geer)

CHAPTER FOUR: U.S. AIR FORCE STRATEGIC F-111s

68-0278 (B1-50)
68-0278 was delivered to the USAF on December 31, 1970. It is seen here at Offutt AFB in September 1980. While assigned to the 380th Bomb Wing at Plattsburgh AFB, the aircraft carried the nickname and nose art *A Wing An'10 Prayers* and was named after 380th Bomb Group WWII B-24 (44-42378). The aircraft was converted to an F-111G and sold to Australia. A8-278. (George Cockle)

68-0279 (B1-51)
68-0279 was delivered to the USAF on December 31, 1970. It is seen here on May 20, 1978. While assigned to the 509th Bomb Wing, the aircraft crashed and was destroyed on July 30, 1980 in Canada. The aircraft experienced an uncommanded left roll during an Automatic TFR letdown. The crew was unable to gain control and was forced to eject. The aircraft had logged 707 flights and 2,562.4 hours when it crashed. (Author's Collection)

68-0280 (B1-52)
68-0280 was delivered to the USAF on January 31, 1971. It is seen here in August 1971. While assigned to the 380th Bomb Wing at Plattsburgh AFB, the aircraft collided with 69-6505 during a night formation rejoin in Vermont on February 3, 1975. Following successful ejections by both crews, both aircraft crashed. The aircraft had logged 287 flights and 1,077.8 flight hours. (Terry Love Collection)

68-0281 (B1-53)
68-0281 was delivered to the USAF on December 31, 1970. It is seen here at Carswell AFB on November 4, 1984. The aircraft was converted to an F-111G and later sold to Australia as A8-281. (Brian C. Rogers)

68-0282 (B1-54)
68-0282 was delivered to the USAF on January 27, 1971. It is seen here at Offutt AFB in April 1981. While assigned to the 380th Bomb Wing at Plattsburgh AFB, the aircraft carried the nickname and nose art *Dream Gal* FB-111A. It won at least five Fairchild Trophies (SAC Bombing Competition). The aircraft was converted to an F-111G and sold to Australia as A8-282. (George Cockle)

68-0283 (B1-55) NO PHOTO AVAILABLE
68-0283 was delivered to the USAF on paper on March 26, 1971. The aircraft was never physically delivered. It had crashed and was destroyed near Manderville, Louisiana, on January 8, 1971, during its acceptance flight. The cause of the crash is unknown. The crew ejected out of the envelope. All the ejection module systems appeared to have worked until it hit the ground at too high a rate of descent. The crew, Lt Col Bruce Stocks and Major Billy Gentry, was found dead in the capsule. The aircraft was brand new having logged only 4 flights and 12.0 hours when it crashed.

68-0284 (B1-56)
Right: 68-0284 was delivered to the USAF on January 29, 1971. It is seen here at Offutt AFB in September 1982. The aircraft was assigned to the 509th Bomb Wing, Pease AFB, New Hampshire. It carried the nickname and nose gear door art *Next Objective*. 68-0284 was delivered to Barksdale AFB on July 10, 1991. It is now on display at the 8th Air Force Museum there. (George Cockle)

68-0285 (B1-57) NO PHOTO AVAILABLE
68-0285 was delivered to the USAF on February 27, 1971. While assigned to the 380th Bomb Wing, the aircraft crashed and was destroyed on October 28, 1977, on the Ashland Range, Maine. The aircraft crashed after a loss of hydraulics caused by an afterburner fire. The crew ejected successfully with no injuries. When it was lost, 68-0285 had logged 384 flights and 1,391.7 flight hours.

CHAPTER FOUR: U.S. AIR FORCE STRATEGIC F-111s

68-0286 (B1-58)
68-0286 was delivered to the USAF on January 30, 1971. It is seen here at Nellis AFB in August 1976. The aircraft was assigned to the 380th Bomb Wing and carried the nicknames *Miss Giving* and *SAC Time*. 68-0286 was retired in June 1991, and is now on display at Plattsburgh AFB. (Author)

68-0287 (B1-59)
68-0287 was delivered to the USAF on February 16, 1971. It is seen here in September 1984. The aircraft was assigned to the 509th Bomb Wing and named *Liberator II*. It was retired and transferred to Lowry AFB, Colorado as stores loading trainer on November 9, 1990. With the closing of Lowry, it moved to Wings Over the Rockies Air and Space Museum, Denver, Colorado. (Tom Kaminski Collection)

68-0288 (B1-60)
68-0288 was delivered to the USAF on February 12, 1971. It was assigned to the 380th Bomb Wing and carried the nickname and nose art *Peace Offering*. 68-0288 is seen here at AMARC shortly after its arrival on July 3, 1991. When retired, the aircraft had a total of 5,685.4 flight hours. (Bob Shane)

68-0289 (B1-61)
68-0289 was delivered to the USAF on February 23, 1971. It was assigned to the 380th Bomb Wing. It is seen here in September 1988, with the nickname and nose art *Queen Hi*. 68-0289 was converted to an F-111G. (Author's Collection)

68-0290 (B1-62)
68-0290 was delivered to the USAF on February 22, 1971. It is seen here at Nellis AFB in February 1975. The aircraft was assigned to the 380th Bomb Wing. 69-0290 crashed and was destroyed on December 23, 1975, near Loring, Maine. The crash resulted from a massive engine explosion and fire. The crew ejected successfully with no injuries. When lost, 68-0290 had logged 348 flights and 1,287.7 flight hours. (Author)

68-0291 (B1-63)
68-0291 was delivered to the USAF on June 30, 1971. It is seen here at Nellis AFB in August 1981. The aircraft was assigned to the 380th Bomb Wing and carried the nickname and nose art *Shady Lady*. 68-0291 was converted to an F-111G and later sold to Australia as A8- 291. (Terry Love Collection)

CHAPTER FOUR: U.S. AIR FORCE STRATEGIC F-111s

68-0292 (B1-64)
68-0292 was delivered to the USAF on February 27, 1971. It is seen here on November 21, 1971, at McCoy AFB, Florida, marked with the winged 2 of the 2nd Air Force. While assigned to the 380th Bomb Wing at Plattsburgh AFB, the aircraft carried the nickname and nose art *Liberty Belle*. The aircraft arrived at AMARC on June 6, 1991. When retired, the aircraft had a total of 5,914.6 flight hours. (Tom Brewer Collection)

69-6503 (B1-65)
69-6503 was delivered to the USAF on February 25, 1971. It is seen here at Pease AFB, New Hampshire on July 5, 1987, with the tiger stripe painted rudder carried on some aircraft of the 393rd Bomb Squadron Tigers. The aircraft was assigned to the 509th Bomb Wing, Pease AFB, New Hampshire. 69-6502 carried the nickname and nose art *Straight Flush*. The aircraft was converted to an F-111G. (F. A. Jackson)

69-6504 (B1-66)
69-6504 was delivered to the USAF on February 28, 1971. It is seen here at Ellsworth AFB in September 1988. The aircraft was assigned to the 380th Bomb Wing at Plattsburgh AFB. 69-6504 was converted to an F-111G. (Author)

69-6505 (B1-67) NO PHOTO AVAILABLE
69-6505 was delivered to the USAF on February 28, 1971. While assigned to the 380th Bomb Wing at Plattsburgh AFB, the aircraft collided with 68-0280 during a night formation rejoin over Vermont, on February 3, 1975. Following successful ejections by both crews, both aircraft crashed. 69-6505 had logged 339 flights and 1,304.3 flight hours at the time of the crash.

69-6506 (B1-68)
69-6506 was delivered to the USAF on March 16, 1971. It is seen here at Forbes Field ANGB, Kansas in December 1978. The aircraft was assigned to the 509th Bomb Wing, Pease AFB, New Hampshire, and named *Full House*. The aircraft was converted to an F-111G and later sold to Australia as A8-506. (Jerry Geer)

69-6507 (B1-69)
69-6507 was delivered to the USAF on March 22, 1971. It is seen here at Offutt AFB in September 1980. It carried the nickname *Sleepy Time Gal* while assigned to the 509th Bomb Wing. While assigned to the 380th Bomb Wing, 69-6507 carried the nickname and nose art *Madame Queen*. It was retired on June 10, 1991, and is now on display at Castle AFB Museum. (George Cockle)

69-6508 (B1-70)
69-6508 was delivered to the USAF on March 31, 1971. It is seen here at Offutt AFB in March 1981. While assigned to the 380th Bomb Wing, 69-6508 was damaged on September 29, 1972, when it made a landing at the wrong air field – Clinton County, New York, and ran off the runway. It was repaired and returned to service. While assigned to the 509th Bomb Wing, it carried the nickname and nose gear door art *Strange Cargo*. The aircraft was converted to an F-111G. (George Cockle)

CHAPTER FOUR: U.S. AIR FORCE STRATEGIC F-111s

69-6509 (B1-71)
Right: 69-6509 was delivered to the USAF on March 24, 1971. It is seen here on August 6, 1983 marked as the 509th Bomb Wing Flagship *Spirit of the Sea Coast*. The aircraft was assigned to the 509th Bomb Wing, Pease AFB, New Hampshire, and was nicknamed *Spirit of the Sea Coast* and *Max Effort*. It was retired on July 12, 1991, and is now on display at Whiteman AFB, Missouri, the new home of the 509th Bomb Wing. (Author's Collection)

69-6510 (B1-72)
Below: 69-6510 was delivered to the USAF on March 30, 1971. It is seen here at Carswell AFB on June 6, 1986, carrying the nickname and nose art *Sleepy Time Gal*. It was assigned to the 380th BW. The aircraft was converted to an F-111G. (Brian C. Rogers)

69-6511 (B1-73) NO PHOTO AVAILABLE
69-6511 was delivered to the USAF on March 16, 1971. While assigned to the 380th Bomb Wing at Plattsburgh AFB, 69-6511 crashed and was destroyed on June 7, 1976. The aircraft crashed in Vermont after departing controlled flight due to a flight control problem (roll couple departure) during an enroute descent. The crew ejected successfully. When lost, 69-6511 had logged 456 flights and 1,763.5 flight hours.

69-6512 (B1-74)
69-6512 was delivered to the USAF on April 16, 1971. It is seen here at Carswell AFB on March 27, 1987, carrying the nickname and nose art *Royal Flush*. It was assigned to the 380th Bomb Wing at Plattsburgh AFB. The aircraft was converted to an F-111G and was sold to Australia as A8-512. (Brian C. Rogers)

69-6513 (B1-75)
69-6513 was delivered to the USAF on April 30, 1971. It is seen here at McConnell AFB in October 1989. While assigned to the 509th Bomb Wing, 69-6513 carried the nickname and nose art *Top Secret*. The aircraft arrived at AMARC on July 1, 1991. When retired, the aircraft had a total of 6,059.0 flight hours. (Author)

69-6514 (B1-76)
69-6514 was delivered to the USAF on May 28, 1971. It is seen here at Pease AFB on July 5, 1987. The aircraft was assigned to the 509th Bomb Wing, Pease AFB, New Hampshire. It carried the nickname and nose art *Double Trouble*. The aircraft was converted to an F-111G and was sold to Australia as A8-514. (F. A. Jackson)

FB-111A Names and Nose Art

During 1986, the 380th Bomb Wing adopted some of the 380th Bombardment Group's B-24 nose-art and started to name each aircraft assigned to the Wing. Nose-art was painted on the right side below the cockpit and the name repeated on the nose wheel doors with the crew members and Crew chief's names.

The 509th Bomb Wing traces its origins from the 509th Composite Group equipped with B-29s during World War II. Its FB-111As had nose-art painted on the nose wheel doors with details of the original a/c stenciled beside it. All the FB-111As had names, but not all carried nose-art associated with the names. As with the other SAC bombers (B-1B and B-52G/H), the same nose-art and name appeared on different aircraft as they were transferred between units or sent to the depot for maintenance/modification. The following pages contain some of the nose art from the 380th and 509th Bomb Wings.

CHAPTER FOUR: U.S. AIR FORCE STRATEGIC F-111s

67-0161 – LIQUIDATOR – 380th Bomb Wing (Craig Kaston Collection)

67-0163 – MOONLIGHT MAID – 380th Bomb Wing (Author)

67-7192 – SLIGHTLY DANGEROUS – 380th Bomb Wing (Craig Kaston Collection)

67-7193 – TIGER LIL – 509th Bomb Wing (Chris McWilliams)

67-7194 – YANKEE CLIPPER – 509th Bomb Wing (Chris McWilliams)

67-7195 – DAVE'S DREAM – 509th Bomb Wing (Chris McWilliams)

67-7195 – BIG STINK – 509th Bomb Wing (Chris McWilliams)

67-7196 – RUPTURED DUCK – 509th Bomb Wing (Chris McWilliams)

68-0239 – ROUGH NIGHT – 380th Bomb Wing (Author's Collection)

68-0241 – UNDECIDED – 380th Bomb Wing (Craig Kaston Collection)

68-0244 – LUCKY STRIKE (Right Side) – 380th Bomb Wing (Craig Kaston Collection)

68-0244 – 380th BOMBARDMENT GROUP ASSOCIATION (Left Side) – 380th Bomb Wing (Craig Kaston Collection)

GENERAL DYNAMICS F-111 AARDVARK

68-0245 – REDDY TEDDY – 380th Bomb Wing (Author's Collection)

68-0248 – FREE FOR ALL – 380th Bomb Wing (Brian C. Rogers)

68-0249 – LITTLE JOE – 380th Bomb Wing (Author's Collection)

68-0249 – SPIRIT OF THE SEACOAST – 509th Bomb Wing (Chris McWilliams)

68-0250 – SILVER LADY – 380th Bomb Wing (Bob Shane)

68-0251 – SHY-CHI BABY – 380th Bomb Wing (Author's Collection)

68-0255 – SLEEPY TIME GAL – 380th Bomb Wing (Brian C. Rogers)

68-0257 – NEXT OBJECTIVE – 509th Bomb Wing (Chris McWilliams)

68-0258 – THE WILD HARE – 509th Bomb Wing (Chris McWilliams)

68-0259 – NECESSARY EVIL – 509th Bomb Wing (Chris McWilliams)

68-0265 – ANGEL IN DE SKIES – 380th Bomb Wing (Douglas Slowiak/Vortex Photo Graphics)

68-0265 – LIBERATOR II – 509th Bomb Wing (Chris McWilliams)

CHAPTER FOUR: U.S. AIR FORCE STRATEGIC F-111s

68-0267 – MEMPHIS BELLE II (Right Side) – 509th Bomb Wing (Chris McWilliams)

68-0267 – MEMPHIS BELLE II (Left Side) – 509th Bomb Wing (Chris McWilliams)

68-0269 – SAD SACK – 380th Bomb Wing (Craig Kaston Collection)

68-0271 – BOMERANG – 509th Bomb Wing (Chris McWilliams)

68-0272 – UP AN ATOM – 509th Bomb Wing (Chris McWilliams)

68-0273 – MILK WAGON – 509th Bomb Wing (Chris McWilliams)

68-0274 – LAGGIN DRAGON – 509th Bomb Wing (Chris McWilliams)

68-0275 – BOMBLE BEE – 509th Bomb Wing (Author's Collection)

68-0276 – GRUESOME GOOSE – 509th Bomb Wing (Chris McWilliams)

68-0278 – A WING AND 10 PRAYERS – 380th Bomb Wing (Jim Benson)

68-0281 – EAGLE ONE – 509th Bomb Wing (Chris McWilliams)

68-0284 – NEXT OBJECTIVE – 509th Bomb Wing (Chris McWilliams)

GENERAL DYNAMICS F-111 AARDVARK

68-0288 – PEACE OFFERING – 380th Bomb Wing (Craig Kaston Collection)

68-0289 – QUEEN HI – 380th Bomb Wing (Craig Kaston Collection)

69-6503 – STRAIGHT FLUSH – 509th Bomb Wing (Chris McWilliams)

69-6506 – FULL HOUSE – 509th Bomb Wing (Chris McWilliams)

69-6507 – SLEEPY TIME GAL – 509th Bomb Wing (Chris McWilliams)

69-6507 – MADAME QUEEN – 380th Bomb Wing (Author's Collection)

69-6508 – STRANGE CARGO – 509th Bomb Wing (Chris McWilliams)

69-6509 – SPIRIT OF THE SEACOAST – 509th Bomb Wing (Chris McWilliams)

69-6509 – MAX EFFORT – SPIRIT OF THE SEACOAST – 509th Bomb Wing (Chris McWilliams)

69-6510 – SLEEPY TIME GAL – 380th Bomb Wing (Brian C. Rogers)

69-6513 – TOP SECRET – 509th Bomb Wing (Chris McWilliams)

69-6514 – DOUBLE TROUBLE – 509th Bomb Wing (Chris McWilliams)

CHAPTER FOUR: U.S. AIR FORCE STRATEGIC F-111s

OTHER F-111 STRATEGIC BOMBER DESIGNS

The Stretched FB-111

The USAF studies continued after development of the FB-111A. The DoD Joint Strategic Bomber Study examined the B-1 and three aircraft alternatives: a stretched FB-111; a re-engined B-52 (designated B-52I or J); and a stand-off cruise missile launching aircraft based on the Boeing 747. To evaluate the alternative aircraft, they were compared in three critical phases of a strategic bombing mission. These phases were launch, flight to the target area, and penetration to the target itself. To be effective, a strategic weapon system must do well in all three phases. The following conclusions were presented to the Senate Armed Services Committee by USAF Secretary McLucas on April 17, 1975.

FB-111G

The first alternative, the stretched FB-111G, did well in launch survivability, but was deficient in the second phase — flight to the target area. The problem with the FB-111G was that the basic airframe was too small, its physical size was about half that of the B-1, and its weight was about one-third that of the B-1. The stretched FB-111G design was range-payload limited, that is, it did not have the space to carry sufficient fuel or weapons to give adequate target coverage.

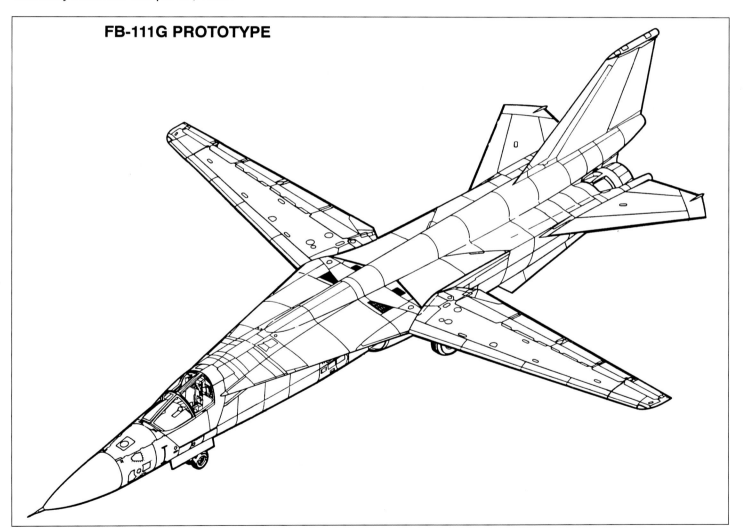

FB-111G PROTOTYPE

FB-111H

General Dynamics recognized the limitations of the small physical size of the basic airframe and proposed a new stretched-stretched version, called the FB-111H (or GD-916). This new version was still lacking, in that it could only carry half the number of weapons as the B-1, and, with a full weapons load, was completely dependent on air refueling tankers; fully loaded with weapons, it couldn't cross the Atlantic Ocean without air refueling. In comparison with the B-1 design, the FB-111H, even when completely fueled at start of a standard penetration, could not reach targets around Moscow with a full weapons load and still be able to return to a recovery base.

The smaller airframe of the stretched FB-111 also could not carry enough avionics and electronic countermeasures to defeat the projected enemy defenses. The DoD study showed the stretched FB-111 would suffer heavy losses throughout the penetration phase by the Soviet defenses which were expected to be in place in the 1980s.

Size also affected the growth potential. The study commented that a major reason the life of the B-52 had been able to be extended to over 30 years of operation was the fact that the design was large enough to accept newly developed avionics and ECM equipment. The size of the stretched FB-111 limited the growth potential.

In the DoD study, the stretched FB-111 was far less than half as cost effective as the B-1. The DoD analyses showed that to achieve comparable target effectiveness, it would take ten times as many FB-111Hs as B-1s, many more tankers, and additional bases. Such an FB-111 force would be cost prohibitive.

In summary, the DoD study found that the stretched FB-111, because of its shortcomings in range-payload, ECM, and growth potential, would be the least cost effective alternative. Both the DoD and the GAO concluded the proposed stretched FB-111 was "clearly non-competitive" with the B-1 design. The decision was made to pursue development of the B-1 over the stretched FB-111.

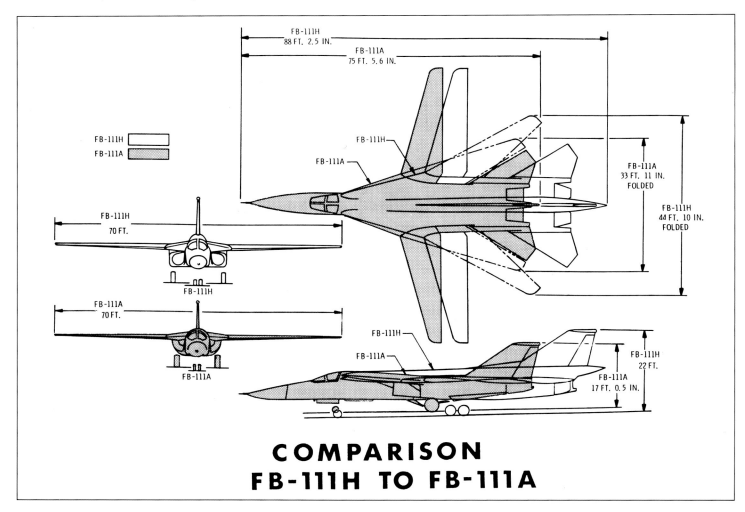

COMPARISON FB-111H TO FB-111A

CHAPTER FOUR: U.S. AIR FORCE STRATEGIC F-111s

Long Range Combat Aircraft (LRCA)

Late in 1979, the DoD initiated another study through the Air Force Scientific Advisory Board to determine the direction that future strategic bomber development should take. It concluded that the nation's next strategic bomber should have multi-mission capability, rather than a single dedicated role, and would have to provide Initial Operational Capability (IOC) in 1987. Three aircraft proposals options were reviewed to fulfill these requirements. Rockwell International's proposal was a modified version of the B-1A. General Dynamics proposal was for the FB-111B/C (which was quite similar to the FB-111H design which had been ruled out as a strategic bomber by the Joint Strategic Bomber Study in 1975). This new FB-111B/C was to have been a stretched version, based on modifying 66 FB-111A and 89 F-111D aircraft to extend the range and increase the aircraft payload. The third option was an all new design, Northrop's Advanced Technology Bomber (ATB) commonly known as the Stealth Bomber, later designated the B-2.

The DoD was also reviewing the issue of whether a single aircraft could meet all of the objectives, and if not, could they be satisfied by producing a multi-role bomber first, with the

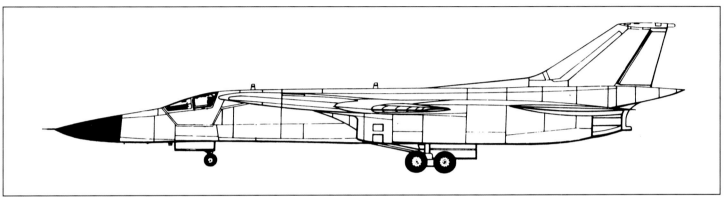

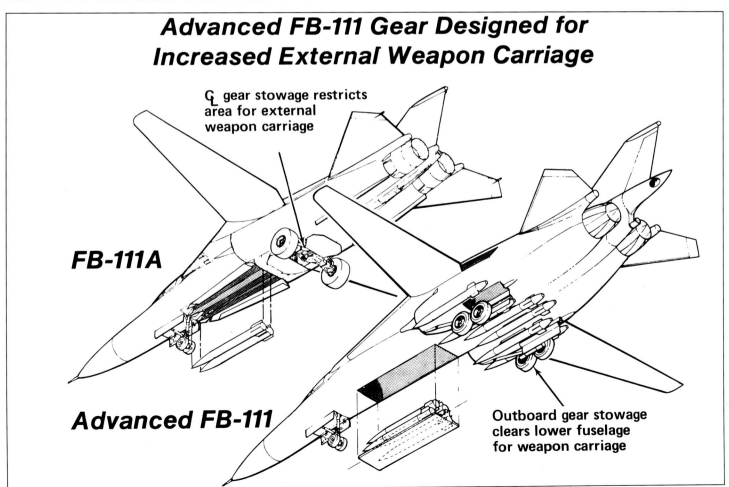

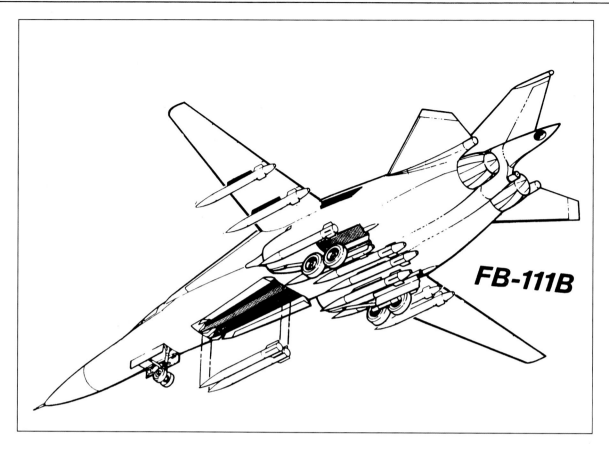

FB-111B

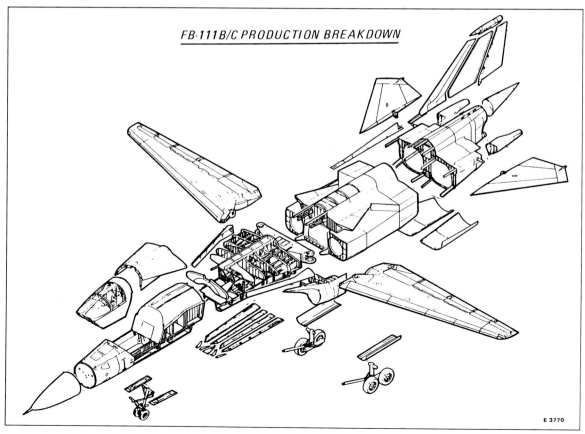

FB-111B/C PRODUCTION BREAKDOWN

ATB phased in at a later date. The study focused on bomber payload and range, penetration capability, multi-role application, flexibility, cost, initial operational capability and the maturity of the technology.

Creating the FB-111B/C would entail stretching the fuselage from 73 feet to 88 feet. This would have given the new version increased weapons capability, with extended range and payload. An additional 31,822 pounds of fuel would have been added, bringing the total fuel capacity to 64,574 pounds. The aircraft would have been equipped with new defensive avionics as a retrofit after delivery to meet the IOC date. The aircraft was to be re-engined with two F101 engines. An auxiliary power unit was required and would have been added in the tail cone. The main landing gear was to have been moved outboard to allow for increased gross weight and to increase the weapons carrying capability. The weapons load would have been increased from the existing capacity of six weapons (two in the weapons bay, and one each on four underwing pylons) to six external nuclear weapons and either four SRAMs, five B61 nuclear bombs, four B57 nuclear bombs, or three B43 nuclear bombs in the weapons bay. The new design could not carry ALCMs internally. Carrying the ALCM externally would increase drag, severely decreasing the aircraft range. Also, ALCMs carried on wing stations were a nuclear safety concern because they would be dangerously close to the ground at takeoff and landing attitudes. The stretched FB-111 would also depend heavily on tanker support, even with the increased fuel capacity brought about by a longer fuselage.

The changes to the B-1A to fulfill the LRCA requirements actually reduced the complexity of the aircraft. These changes included simplification of the engine inlet systems and the overwing fairings. The B-1 version of the LRCA could carry 14 ALCMs externally and eight ALCMs internally. The aircraft design takeoff weight would be 477,000 pounds. The aft stores bay would still be available to carry eight gravity nuclear bombs, eight SRAMs, or an auxiliary fuel tank. Both the B-1 LRCA and the stretched FB-111 would use radar absorbing materials (RAM) to reduce their radar cross-section (RCS) by a factor of 10.

The two candidates – the FB-111B/C and the B-1 LRCA – appeared to be able to meet the 1987 IOC date, while the ATB would present at best a high risk development program. The modified B-1 and the stretched FB-111 presented different issues which affected the DoD confidence in reaching the 1987 IOC. General Dynamics had a production base in place producing the F-16; however, extensive modifications would be required to convert the F-111D and FB-111A. The entire mid and aft fuselage sections would be new, as would the engines. An extensive test program would be necessary to validate design and determine the performance of the stretched aircraft. The B-1A prototypes were still flying, and while the modifications required were substantial, they were relatively straightforward.

This final study by the USAF and DoD led to selection of the B-1B as the next strategic bomber, and in October 1981 President Ronald Reagan announced that the USAF would acquire 100 B-1B aircraft. The ATB program was also continued, with plans to eventually procure 150 B-2 stealth bomber aircraft.

ADVANCED FB-111 CHARACTERISTICS

	FB-111A	FB-111(H) – FB-111B/C
DIMENSIONS		
Wing Span		
Spread	70 feet	70 feet
Fully Swept	33 feet, 11 inches	44 feet 10 inches
Wing Sweep	16 to 72 degrees	16 to 60 degrees
Length	73 feet 6 inches	88 feet 2.5 inches
Height	17 feet 1.4 inches	22 feet
WEIGHT		
Weight Empty	47,445 pounds	51,832 pounds
Max Takeoff Weight	116,115 pounds	140,000 pounds
Max Inflight	122,900 pounds	155,500 pounds
Fuel Capacity	32,000 pounds	64,000 pounds
PERFORMANCE		
Max Speed		
Altitude	Mach 2.2	Mach 2.0 plus
Sea Level	Mach 1.1	High Transonic
Service Ceiling	Above 50,000 feet	Above 50,000 feet
Takeoff Distance	7,400 feet	6,650 feet
Landing Distance	5,300 nautical miles 1,200 nautical mile low level high speed dash	Twice the sea level dash range of the FB-111A with increased payload
ENGINES	Two Pratt & Whitney TF30-P7 engines	Two General Electric F101 GE-100 engines
Bypass Ratio	0.73	2.01
Thrust Class	20,000 pounds	30,000 pounds
LANDING GEAR		
Nose	Twin wheel nose unit retracts forward	SAME
Main Gear	One unit, gear retracts Wheels are stored side by side in the fuselage between engine air intakes	New, simplified tandem gear retracts aft, allowing weapons to be carried on the fuselage
AUXILIARY POWER UNIT	None	APU Provides simultaneous starting of both engines and provides power for ground options.
WEAPONS BAY	Two Nuclear Weapons	H Model – Up to Six Nuclear Weapons B/C Models – Two Nuclear Weapons

CHAPTER FIVE

U.S. Navy F-111s

F-111B

During the late 1950s, the US Navy was looking for a new aircraft for fleet defense. The aircraft proposed for this mission would be tied to a new missile the AAM-N-10 Eagle. Rather than a sleek Mach 2+ aircraft, the design proposed for this mission was the straight-winged blunt-nosed Douglas Missileer. The Douglas Missileer was developed as a flying launch pad instead of a combat machine. Its sole purpose in being was to launch six Eagle long-range air-to-air missiles in the direction of a distant attacker. The missiles themselves were considered the interceptors.

Development contracts for the Eagle missile and its airborne missile control system were awarded by the Navy on December 5, 1958. Bendix was to develop a long-range interceptor missile for fleet defense, a type now known as "Stand-off" missiles. On July 21, 1961, Douglas won a lengthy competition to design an airframe around the Bendix system and build two prototypes as the XF6D-1. The AAM-N-10 Eagle was developed by Bendix with the construction of the airframe to be carried out by Grumman. Six of these weapons were to be carried under the wings of the Missileer. On a typical operation scenario, the F6D-1 launched from a carrier and flew to a patrol zone about 150 miles from the carrier. It was designed to loiter at an altitude of 35,000 feet for ten hours. If a missile launch was necessary, after the Missileer acquired the target, the missile would launch and accelerate to Mach 4, climbing to 100,000 feet seeking its target. A lock-on cruise and homing radar guidance system directed the Eagle to its target – up to 100 miles from the launch point. The Eagle missile was 20 feet long and weighed approximately 2,000 pounds. It could carry either conventional or nuclear warheads.

The F6D-1 Missileer design had a wingspan of 70 feet with an area of 630 square feet. Length was 53 feet and the height 18 feet 1 inch. The tip of the fin could be folded to reduce the height to 16 feet 3 inches for stowage below decks. Design gross weight was 60,000 pounds. Two TF30-P-2 turbofans by Pratt & Whitney were to power the Missileer with a total of 16,500 lbs. thrust. This would give a maximum speed of 543 mph. The F6D-1 was to be operated by a crew of two.

The Douglas Company proceeded with the design and development of the Missileer and received an order for 120 F6D-1s worth about six million dollars. Due to a reevaluation of the "Stand-off" concept, the contracts were withdrawn on April 25, 1961, before any prototype construction had begun.

Robert McNamara taking over, in January 1961, as John Kennedy's Secretary of Defense strongly backed the TFX. McNamara's TFX was a unique concept; for the first time, a single airplane was to be designed from the beginning for both the Navy and the Air Force. Under the TFX program, F-111B development, like that of the USAF F-111A, reflected Secretary McNamara's September belief that each service's long-range requirements could be met with one aircraft. The bi-service F-111 would replace the F-105 for the USAF. It was to replace the carrier-based F4H (F-4 Phantom). This eliminated the Navy's chances for getting the F6D Missileer as the F4H's replacement. The F4H, which topped all Navy interceptors in speed, altitude, and range, was introduced into the Fleet in January 1961, only a few months before the rejection of the single mission Missileer interceptor. This concept appeared to be practical and economically sound. But it ignored the wide differences between the operational requirements of the two services. When first presented with the TFX concept, both the USAF and US Navy rejected the plan. They felt it was physically impossible to build a common airplane for such diverse missions.

Despite the protests, the DoD went ahead with the requirements for the TFX (Tactical Fighter-Experimental). The Navy plane, designated F-111B, would have long wings for slow carrier approaches, and a short nose to allow for a larger radar dish, better vision over the nose during landing approach, and ease of stowage in the confined space of an aircraft carrier. The two versions would share 84 percent of the overall structure including the variable-sweep wings. The F-111B would be equipped with the AN/AWG-9 to control and

CHAPTER FIVE: U.S. NAVY F-111s

fire the AIM-54/A Phoenix missile. Design, development manufacturing, final assembly and delivery of the F-111B was to be accomplished by the Grumman Aircraft Corporation. The USAF authorized General Dynamics to negotiate the subcontract in September, two months before its official ratification.

The first F-111B flew on May 18, 1965, and was powered by the initial TF30-P-1 engine. It flew for one hour and 18 minutes after taking off from Grumman's Peconic, New York, facility. It had rolled out of the Grumman's Bethpage, New York plant seven days earlier. The USAF immediately accepted for the Navy the first YF-111B, sending it to the Patuxent Naval Air Test Center in Maryland where all F-111Bs would be tested. The F-111B's Phoenix missile system would undergo tests in California, at the Hughes, Culver City Plant and at the Naval Point Mugu/Pacific Missile Test Center. The aircraft reached supersonic speed on July 1, 1965.

The first flight of the RDT&E F-111B's took place in May 1966. The F-111B development took longer than the F-111A's because of difficulty in integrating the new Phoenix missile system with the aircraft. The F-111B's first successful launch of the AIM-54A Phoenix took another six months. The F-111B also shared the F-111A's engine problem. The Navy believed these would be solved with the P-12 (the Navy version of the TF30), which was scheduled to equip F-111B production aircraft and would retrofit the RDT&E F-111Bs, beginning in late 1966.

The F-111's crew module lacked sufficient forward visibility for a carrier-based aircraft. On March 11, 1967, the design of a new module for the F-111B was authorized, even though this would mean aerodynamic changes and more differences between the F-111A and F-111B. Redesign of the fuselage to fit the P-12 engine had caused the overall percentage of common parts, once around 80, to fall below 70. Redesign of the F-111B's crew module (including pilot elevation and increased windshield slant) was a further change. F-111B aircraft 6 and 7 were built with this new production design.

Meanwhile, continued USAF and USN efforts to check F-111 weight increases proved futile. The first F-111B proto-

type flown (modified F-111A), weighed 69,000 pounds, too much to permit the aircraft's operation from carriers smaller than the USS Forrestal. The first F-111B production aircraft (due to fly in 1968) would weigh 75,000 – about 20,000 pounds more than originally planned.

A definitized contract (for production aircraft) was signed on May 10, 1967 by the USAF. It called for 24 Navy F-111Bs which were included in the 493 F-111s covered by the USAF contract. On June 30, 1968 Grumman delivered the first production aircraft to the USAF, for the Navy.

On July 9, 1968, the USAF stopped work on the F-111B after the House Armed Services Committee joined the Senate in disapproving a $460 million appropriation requested by the Defense Department for further development and procurement of 30 aircraft. In August 1968, the F-111B production was canceled. Cancellation of the Navy F-111B led General Dynamics to end its relationship with Grumman and Hughes. Hughes, as associate contractor under the Navy contract, continued to develop the Phoenix missile system for use by the Navy in the VFX. The Navy's withdrawal, along with the British government's cancellation of its F-111K pur-

F-111Bs 152715, 151972, and 152714 are seen here during Phoenix missile testing, parked on the Hughes flight line in Culver City, California. The different shape of the preproduction nose and cockpit area on 151972 (the middle aircraft) is apparent. (Marty Isham Collection)

253

F-111B number 3 (151972) is seen here during a AIM-54A Phoenix separation test. (Mick Roth Collection)

chase, forced the USAF to adjust its plans for procurement of the F-111. By fiscal year 1970, the May 1967, contract's buy of 493 F-111s over four years had been stretched out to six years.

On February 28, 1969, with delivery of the seventh and last F-111B production ended. A total of seven aircraft were accepted, five RDT&E F-111Bs and two production aircraft. Two F-111Bs were lost in crashes and a third was severely damaged in landing. The Navy used the remaining four to continue testing the Phoenix missile system and P-12 engine. Both would be used in the F-111B's successor, the VFX. The VFX (Fixed Wing Fighter Experimental) program authorized for development by Congress in July 1968 culminated in the development and production of the Grumman F-14.

F-111B AIRCRAFT DESCRIPTION

The aircraft had a crew of two, the Pilot and the Missile Control Officer (MCO), seated in a side-by-side arrangement. The first three aircraft used individual ejection seats for the crew members. The last four and the planned production aircraft had an ejectable crew module similar to the USAF version. The shorter fuselage nose radome contained the long-range panoramic radar for interceptor role. The longer wing provided improved low-speed ferry and loiter performance. The F-111B's overall length of 66 feet and 9 inches was about 6 feet under the F-111A's; its 70-foot wing span (same as the FB-111A) was 7 feet longer than the F-111A's. It had an enlarged ventral fin and the lower fuselage between the engines housed a carrier arrester hook. The P-12 engine (another version of the TF30) with a maximum thrust of 20,250 pounds in afterburner – 1,700 pounds more than the F-111A's P-3. It carried six AIM-54A Phoenix air-to-air missiles, developed by Hughes specifically for the Navy. The Phoenix's fire-control system owed much to the USAF ASG-18 system (developed in the early 1960s) for launching nuclear-tipped AIM-47A air-to-air missiles – then known as GAR-9 Falcons, originally meant for the North American F-108 Rapier (canceled by OSD in September 1959).

Two Pratt and Whitney TF30-P-12 afterburning turbofan jet engines, mounted within the fuselage, powered the aircraft. Preventative measures, in the form of vortex destroyers (air nozzles which created an air curtain), were located in the lower surface of the fuselage aft of the hinge lines of the weapons bay doors to prevent foreign object damage (FOD) during static operation and takeoff and landing. The first five aircraft 151970 through 151974 had Triple Plow I inlets, and the last two, 152714 and 152715, had Triple Plow II intakes.

AIRFRAME

The airframe was similar to the early F-111A's, except it had a shorter nose to facilitate handling and storage aboard ship. It had the long wings of the FB-111A and F-111C. The fuselage of the two production F-111Bs (152714 and 152715) was about two feet longer with the added length between the windshield and the radome.

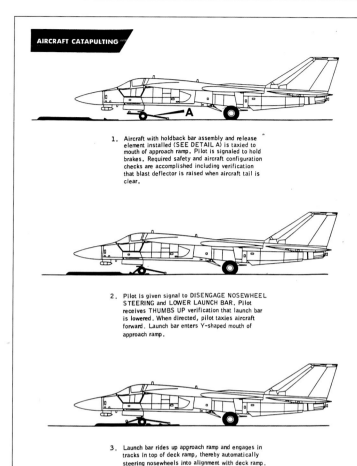

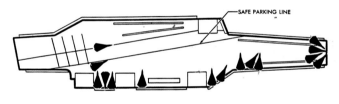

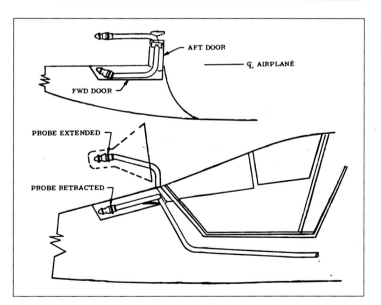

NOSE RADOME

The nose section of the aircraft consisted of a radome which housed the Phoenix missile system antenna. The radome could be folded upward for access to the antenna and associated systems.

LANDING GEAR SYSTEMS

The F-111B had the heavy weight tricycle-type retractable landing gear used on the FB-111A and the F-111C, and an operational tail bumper and arresting hook. The nose landing had an operational launch bar for aircraft catapult operation. The electrically controlled arresting hook system used for carrier deck landings or for emergency arrestment of shore-based aircraft, was forced down by gravity and hydraulic accumulator pressure. The hook was retracted hydraulically. Emergency operation of the arresting hook was accomplished through mechanical controls and dashpot pressure.

AIR REFUELING SYSTEM

The F-111B was designed for use with the U.S. Navy probe and drogue in flight refueling system. With this system it was refuelable using Navy tankers, USAF KC-135s with the drogue installed or KC-10 tankers, and foreign tankers equipped with NATO standard refueling systems. The probe on the F-111B (and F-111K) was similar in shape to the probe used on Grumman A-6s. When extended, the probe was located on the aircraft centerline just in front of the windshield. It folded down into the fuselage on the left side, in front of the pilot.

PHOENIX MISSILE LIQUID COOLING SYSTEM

The Phoenix missile liquid cooling system furnished a flow of liquid coolant at a specified temperature and pressure to the Phoenix missile system. This closed loop system employed an air-to-liquid heat exchanger for temperature control of the liquid coolant. The coolant is circulated by the liquid loop pump package through tubing and flexible lines to the missiles attached to the pivot pylons of the wing and to the electronic equipment in the forward bay.

ARMAMENT SYSTEMS.

The primary armament of the F-111B aircraft was the AIM-54/A Phoenix missile. Aircraft components which support the Phoenix missile and initiate its release are included in the armament system. The two inboard external stations on each wing are connected to the wing sweep mechanism to ensure proper positioning of the stores in relation to the slipstream throughout the wing sweep range. The four pivot pylon stations have provisions for carrying the Phoenix missile.

FIREPOWER CONTROL SYSTEMS.

The firepower control system provides the F-111B aircraft with the capability to launch the Phoenix missile. The following systems are integrated to fulfill this mission: Inertial Navigation System AN/AJN-14, Airborne Missile Control System AN/AWG-9, Tactical Telemetry System AN/AKQ-1, and Electronic Altimeter System AN/APN-167. As its primary mission was intercept of airborne threats, it did not have a TFR system like the other F-111s.

The Inertial Navigation System AN/AJN-16 was an analog inertial system which operated with the Phoenix Missile System and associated systems/equipment peculiar to the F-111B aircraft. The inertial navigation system was a self-contained system which does not require external navigational information from associated systems for its operation. The inertial navigation system provided a continuous display of navigational information to the missile control officer; steering, heading, and altitude information to the flight and autopilot instruments; a method of enroute updating of system data; and the capability of storing up to three different destinations.

The Airborne Missile Control System AN/AWG-9 is composed of the radar subsystem, the infrared subsystem, the control and display units and hand control, and the central digital computer. The Tactical Telemetry System AN/AKQ-1 was used with and controlled by the Phoenix Missile System. It provided accurate missile performance information which was sent from the missile back to the tactical telemetry recorder. Examples of data returned to the aircraft are missile miss distance and fuzing indications.

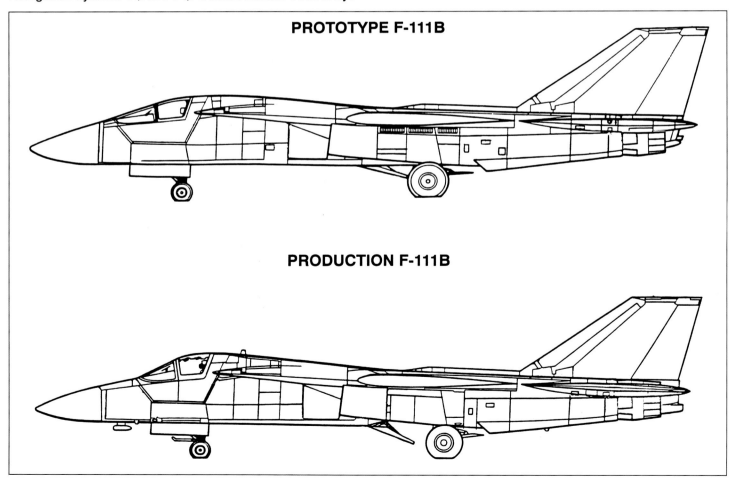

PROTOTYPE F-111B

PRODUCTION F-111B

The Electronics Altimeter AN/APN-167 provided accurate attitude from 0 to 5000 feet above the terrain and also altitude rate of change information. The altimeter contained search and track modes of operation. In the search mode, the system successively examined increments of range with each cycle of operation until the complete altitude range is searched for a ground return signal. In the track mode, the system locked on and tracked a ground return signal giving continuous altitude information.

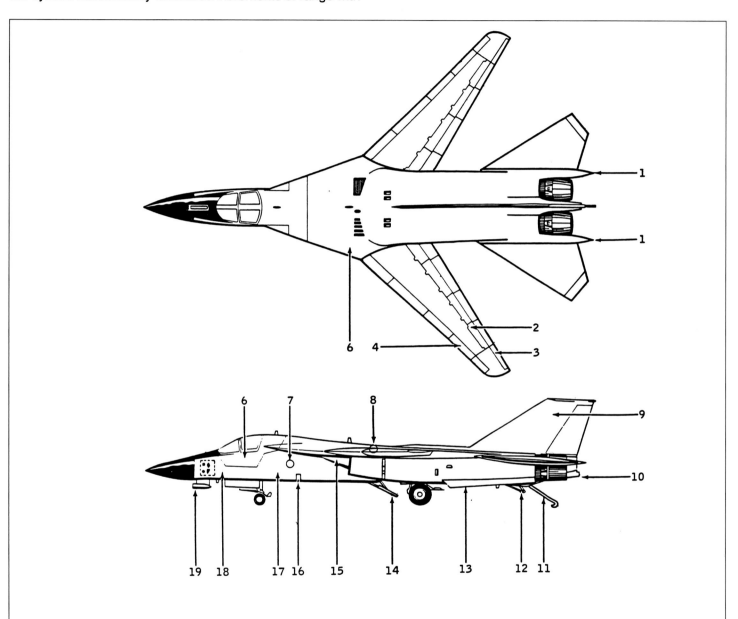

1. SPEED BUMPS
2. SPOILERS
3. FLAPS
4. SLATS
5. ROTATING GLOVE
6. CREW MODULE
7. SINGLE POINT GROUND REFUEL ADAPTER
8. AIR CONDITIONING SYSTEM COOLING AIR INTAKE
9. FUEL VENT TANK
10. FUEL VENT AND DUMP
11. ARRESTING HOOK
12. TAIL BUMPER
13. STRAKE
14. SPEED BRAKE/FORWARD LANDING GEAR DOOR
15. SPIKE
16. FUEL PRECHECK SELECTOR VALVE AND GAGE
17. WEAPONS BAY/AFT ELECTRONIC EQUIP BAY
18. FORWARD ELECTRONIC EQUIPMENT BAY
19. I R DOME

F-111B

151970 (A2-01)
The number one F-111B rolled out of Grumman's Bethpage, New York facility on May 11, 1965, flying for the first time, a week later, on May 18, 1965, and flew supersonic for the first time on July 1, 1965. It reached Mach 2 on March 15, 1968. The first aircraft was used by Grumman for aircraft flight worthiness testing and flew out of NATF Lakehurst, New Jersey. It was fitted with ejection seats. When the test program was completed in July 1968, the aircraft had logged 175 flights and 228.8 flight hours. It was scrapped at NATF Lakehurst in December 1969. (Lockheed Martin Tactical Aircraft Systems)

151971 (A2-02)
Right: The second aircraft made its first flight on October 24, 1965. It was used by Hughes in the testing of the AN/AWG-9 Phoenix missile system, flying from NMC Pt Mugu. It crashed and was destroyed off the California coast (20nm NW of San Miguel Is) on September 11, 1968. It was fitted with ejection seats, but no sign of the crew was found. Killed were Hughes Aircraft test pilot Barton Warren and MCO Anthony Byland. A three by five foot section of the left hand flap washed up on a north shore beach on the island of Oahu, Hawaiian Islands in March 1970, a distance of more than 2700 miles from the crash site. The aircraft had accumulated 121 flights and 172.5 flight hours when it was lost. (Lockheed Martin Tactical Aircraft Systems)

151972 (A2-03)
The aircraft was accepted on December 20, 1965, and flew as part of Grumman's test program from NATF Lakehurst, New Jersey. It was fitted with ejection seats. 151972 successfully launched a Phoenix missile against an airborne target on March 3, 1967, followed on January 18, 1968 by a successful launch from the aircraft while flying supersonic. On October 6, 1969, 151972 was badly damaged when an over wing fairing panel separated during an accelerated windup turn, penetrated the aircraft skin and lodged against the pitch control push-pull rod. The landing was made using only pitch trim. It was repaired at NAS Point Mugu and returned to flying status. It last flew on January 29, 1971, landing at NAS Lakehurst, New Jersey. It had logged 366 flights and 541.4 flight hours. It was scrapped at NATF Lakehurst in December 1971. (Lockheed Martin Tactical Aircraft Systems)

151973 (A2-04)
The number 4 aircraft was used in Grumman's test program and first flew on July 25, 1966. It was the first Super Weight Improvement Program (SWIP) aircraft. 151973 aircraft crashed and was destroyed off the coast of Long Island April 21, 1967. (near Calverton, New York) after double engine failure after takeoff, probably caused by closed engine inlet cowls. With the cowls closed the engines could not get enough air at low speed, and, as a result, flamed out. It was the first F-111B fitted with ejection module. and used as a Phoenix missile testbed. It had recorded 53 flights and 89.1 flight hours when it crashed. (Grumman)

151974 (A2-05)
151974, the second SWIP, was accepted on September 15, 1966, and first flew on November 16, 1966. On July 23 and 24 1968, the aircraft became the first and only F-111 to perform carrier operations. It successfully completed ten arrested landings, ten catapult launches, eight touch-and-go's, and 8 wave-offs. The tests were flown off the USS Coral Sea. Carrier operations occurred after completing arrestor proving tests at PAX River in February 1968. 151974 crash landed at Point Mugu, California on October 11, 1968. Missile Control Officer Bill Bush was injured. Pilot George Marrett was unhurt. On October 10, 1986, it was ferried to NAS Moffett for full scale wind tunnel flight control tests. In February 1970 it was dismantled at Moffett Field. The aircraft had flown 171 flights and logged 271.9 flight hours when retired. (Lockheed Martin Tactical Aircraft Systems)

152714 (A2-06)
The aircraft was close to production configuration with the nose slightly lengthened. It was the first F-111B with TF30 P-12 engines and Triple Plow II intakes. 152714 was accepted on June 30, 1968, and first flew on June 29. It was used by Hughes in missile tests. It was retired to MASDC on July 8, 1969, after flying 34 flights and logging 52.9 hours. It was stricken from the inventory in May 1971. (Lockheed Martin Tactical Aircraft Systems)

152715 (A2-07)
Right: The last F-111B completed, 152715, was accepted on February 28, 1969. It also had TF-30-P-12 engines and Super Plow intakes with double blow-in doors. It was also assigned to Hughes, flying from Hughes Culver City plant, NAS Pt. Mugu, and NAS China Lake. 152715 accomplished three Phoenix system milestones; on March 8, 1969, two Ryan Firebee target drones were shot down at "virtually the same moment", on April 10, 1970, a live phoenix missile downed a QF-9 drone, and on March 18, 1971, two Phoenix missiles were launched simultaneously at two drone aircraft approaching at different angles. It flew its last mission on March 23, 1971, and had logged 234 flights and 322.0 flight hours. The aircraft was retired (struck off) in May, 1971 and stored at China Lake NAS, California. (Gerry Markgraf Collection)

152716 Not Completed
152717 Not completed
153623 to **153642** Not built – Contract Canceled
156971 to **156978** Not built – Contract Canceled

151974 was used for carrier suitability tests. It is seen here prior to touchdown on the deck of the USS Coral Sea, CV-43. (Robert Lawson)

In this view of 151974 in landing configuration, the translating cowls of its Triple Plow I intake are slid forward allowing the additional air required to enter the engines. The tail hook visible between the afterburners is in its stowed position. (Robert Lawson)

CHAPTER SIX

Royal Australian Air Force F-111s

On October 24, 1963, the government of Australia agreed to purchase 24 F-111As. The Australian version was designated F-111C.

The F-111C was a hybrid of the F-111A, the F-111B, and the FB-111A. The F-111C was equipped with eight F-111A type underwing pylons mounted on an F-111B type larger span wing (span of 70 feet when fully extended). It was equipped with an FB-111 type of reinforced landing gear. The 24 F-111Cs were given the USAF serial numbers 67-0125 through 67-0148. Their RAAF serials were A8-125 through A8-148.

The first F-111C was delivered on September 6, 1968. However, the problems with the F-111A's wing carry-through box slipped delivery of the remaining 23 F-111Cs to late 1969. The whole F-111 fleet, including the F-111Cs, were grounded pending verification of their overall structural integrity. The remaining F-111Cs awaiting delivery to Australia were stored at Fort Worth until the structural integrity of the F-111 could be confirmed. In April 1970, a joint agreement deferred Australia's acceptance of the purchased F-111Cs, and specified that the RAAF lease F-4E aircraft until a new wing carry-through box could be installed on all F-111Cs, with the aircraft to be delivered in mint condition. More than a million man hours went into the F-111C modification and refurbishment program started by General Dynamics on April 1, 1972.

In 1973 the F-111C was finally ready for delivery to the RAAF. Australian crews came to the USA. They flew the F-111Cs from Fort Worth to McClellan AFB. As the aircraft were released, Australian crews flew them from the contractor's Convair Aerospace Division in Fort Worth, Texas, to McClellan AFB. Once at McClellan, each F-111C completed between four and six training missions before departure. The first F-111Cs reached Australia on June 1, 1973, replacing the RAAF's Canberra bombers in use since the early 1950s. They were assigned to Nos. 1 and 6 Squadrons based at Amberley, Queensland. The last six of the 24 F-111Cs bought by Australia left the United States on November 26, 1973. This was nearly ten years after the two countries signed a June 1964, F-111 agreement and more than five years since General Dynamics delivered the first F-111C.

The F-111C carries the APQ-113 forward-looking attack radar, which is used for navigation, for air-to-ground ranging and for weapons delivery. This radar can also be used in the air-to-air mode in conjunction with the internal 20-mm cannon or Sidewinder missiles carried underwing. RAAF F-111 crews maintain proficiency in airborne intercepts.

The idea of developing a reconnaissance version of the F-111 originated in July 1966, with six RAAF aircraft and 60 USAF aircraft scheduled for conversion. While the RAAF postponed this plan because of problems with the basic aircraft, the USAF canceled their reconnaissance version altogether. The RAAF didn't lose interest in a reconnaissance F-111, and several years after the first F-111s arrived in Australia, a joint study for a reconnaissance version was made by the RAAF and General Dynamics' Fort Worth Division. The project was approved in July 1977, and the first F-111C to be converted (A8-126) was flown to the U.S. in October 1978. The modifications to A8-126 were completed and RF-111C A8-126 first flew on April 17, 1979. In 1980, the other three aircraft (A8-134, A8-143, and A8-146) were converted by No. 482 Squadron and No.3 Aircraft Depot in Australia.

The RAAF's four RF-111Cs are the only F-111s ever used operationally for photo reconnaissance. Australia's RF-111Cs were operated by No. 6 Squadron Reconnaissance Flight based at RAAF Base Amberley. With the receipt of F-111Gs by the No. 6 Squadron, the RF-111C were transferred to the No. 1 Squadron. The RF-111Cs are all converted F-111C strike aircraft, retaining their strike capability. The RF-111C can conduct long range reconnaissance missions against targets in bad weather conditions and both day and night. This modification, designed by General Dynamics, gave the RF-111C a day/night reconnaissance capability and, with its long range, high speed, and an ability to penetrate at low altitude in all weather conditions, the RF-111C has proven to be a most capable reconnaissance platform. The reconnaissance "kit" is made up of two KS-87C framing cameras, a KA-56E low-

RF-111C A8-134 of the No.6 Squadron, and F-111C A8-127 of the No. 1 Squadron were two of the original 24 F-111Cs purchased by the Royal Australian Air Force. (82nd Wing Photographic, RAAF)

(82nd Wing Photographic, RAAF)

altitude and KA-93A4 high altitude panoramic camera, and an AN/AAD-5 Infrared Linescanner. There is a TV viewfinder which assists with line-up for the photo run.

The aircraft retained the flight envelope and conventional weapon capability of the F-111C strike aircraft, so, it could conduct its missions at low altitudes to evade radar, over a long range, and at high speed and engage targets if necessary. Because of its stability and handling at low altitude, there is no degradation of quality of imagery shot during the mission. A typical low-level mission is flown at 200 feet above ground level with a radius of action more than 1000 nautical miles.

No. 6 Squadron's capabilities with the RF-111C have been proven on numerous occasions. In 1986, 1988 and 1990, the Squadron competed against USAF, U.S. Navy, British Royal Air Force and German Air Force tactical reconnaissance aircraft in Reconnaissance Air Meet (RAM) competitions at Bergstrom Air Force Base, Texas. The No. 6 Squadron did well in all three competitions and won "Best Performance by Top Allied Crew" awards in 1988 and 1990, as well as "Top Crew" and "Top Night Team" awards in 1988. Although RAMs are no longer being held, the Squadron maintains its expertise by participating in Red Flag exercises at Nellis Air Force Base in Nevada, as well as Exercises Pitch Black and Kangaroo and Army exercises in Australia.

The RF-111C received the Avionics Upgrade Project, ensuring its viability as a modern combat system. Thus, the RF-111C aircraft and its support organization are flexible assets that will actively participate in Australia's defense until the year 2020. In order to extend the service life of the F-111C from the year 2010 to 2020, the Government decided to enlarge the RAAF's F-111 fleet and reduce the number of hours being flown on each aircraft per year.

During 1982, to replace F-111Cs which had crashed, Australia took delivery of four used F-111A's (67-0109, 67-0112, 67-0113, and 67-0114). These aircraft were upgraded to F-111C configuration by the RAAF. 67-0106 and 67-0108 were at first included in the buy, but both failed fatigue tests in the U.S. prior to delivery and retired to AMARC.

To upgrade the capability of the F-111C, modern digital avionics were installed as part of the Avionics Update Project (AUP) with the prototype aircraft (A8-132) undergoing ground testing in the U.S.A. The rest of the F-111Cs were modified in Australia with the AUP completed in 1998. The upgrade was engineered by Rockwell at its Palmdale plant. Ninety percent of the electronics in the F-111C were replaced by more capable units. This upgrade should keep the RAAF F-111Cs flying until the year 2020.

To improve the F-111C's effectiveness, the Australian Government decided to incorporate Pave Tack into its F-111Cs. The modification, again designed by General Dynamics, was installed in all F-111C aircraft by the RAAF in the mid-1980s. This new system allowed the F-111C to use laser-guided bombs, a TV-guided bomb called the GBU-15. The AGM-84 Harpoon anti-shipping missile and AGM-142 Have Nap air to ground missile have also been added to make the F-111C a much more effective weapon delivery system than its predecessor.

In order to extend the life-span of the F-111s to the year 2020, the Australian Government bought 15 F-111Gs from the USAF. The F-111Gs have a similar history to the F-111Cs, although they followed the Cs off the production line and were originally flown as FB-111As. They were operated by the USAF and assigned to Strategic Air Command where they had a nuclear strategic role. With the USAF being reduced in size, the FB-111As were modified to enhance their non-nuclear

The RF-111C in the foreground is carrying an AGM-84D Harpoon, while A8-142 in the background has GBU-10s and AIM-9M CATMs on its pylons. (82nd Wing Photographic, RAAF)

weapons delivery capability, assigned to USAF Tactical Air Command and renamed the F-111G. While still belonging to the USAF, the aircraft's avionics systems also went through a modernization program to digitize the old analog systems, and included the weapons delivery, radar, terrain following and communications systems. Because these avionics are a slightly earlier version of those currently being installed in the RAAF's F-111Cs, the two aircraft have a great deal in common. Since they were made at similar times, they also share a common wing set, landing gear and the majority of air frame parts. The F-111G is powered by a Pratt and Whitney P-107 engine which is a later version of the F-111C TF30-P-103 engine and delivers more thrust in full afterburner. The F-111G has approximately 11 percent more power than the F-111C, but slightly decreased range. The engines F-111G are being updated with an Australian hybrid engine called the TF30-P-108. The P-108 has the forward section of the P-109 and the aft section of the P-107. The F-111G's navigation and weapon delivery systems are more accurate than the F-111C's (pre-update) analog systems and significantly more reliable from a maintenance point of view.

While the F-111G is currently not equipped with an infra-red detection and laser designation systems like Pave Tack,(and consequently cannot self-designate precision laser-guided bombs), it can carry such bombs to supplement another airborne or ground designator. It is expected the RAAF will further upgrade the F-111G avionics with a new technology infra-red detection and laser designation system.

A RAAF team went to the United States to select the best 15 F-111Gs being offered for sale by the U.S. Government at optimum cost. The F-111Gs chosen were returned to flying status by the Sacramento Air Logistics Center in a program monitored by the RAAF and then ferried back to Australia in pairs every month; the last three arrived at RAAF Base Amberley in mid-May 1994. The G models come with a logistics package including spare engines and other parts, new digital flight control systems and logistical support including training and facilities. A number of these F-111Gs will replace F-111C aircraft currently in service, and the then-surplus F-111Cs and Gs will be placed in storage. These additional aircraft will be interchanged between storage and active service with the RAAF's current F-111C fleet. The digital flight control systems will be incorporated with an upgrade in the late 1990s.

All the F-111Cs, like A8-138 seen here, were modified to carry the AN/AVQ-26 Pave Tack pods. (Author's Collection)

CHAPTER SIX: ROYAL AUSTRALIAN AIR FORCE F-111s

Fifteen F-111Gs, including this one, were purchased by the Australian government. (82nd Wing Photographic, RAAF)

The RAAF F-111Cs and RF-111Cs have been modified to allow carriage and launch of AGM-84D Harpoon missiles. (Author)

Australian models of the F-111 have been modified to carry the AGM-142 Have Nap missile. This missile, along with the AGM-84D, are also in the USAF inventory and carried on the B-52H. (Author)

F-111C/RF-111C

Twenty-four F-111Cs were built. The factory numbers for the F-111Cs ranged from D1-01 through D1-24. Due to operational problems with USAF F-111s, delivery of the F-111Cs was postponed the second half of 1973. During this time 24 F-4Es were loaned to the RAAF. During 1979 and 1980, four of the F-111Cs were modified to RF-111C configuration. In 1982, Australia took delivery of four USAF F-111As which were upgraded to F-111C configuration. Fifteen F-111Gs were purchased from the U.S. which are to be used to keep the RAAF F-111 fleet operational through 2010. All of the RAAF F-111s are stationed at RAAF Amberley and assigned to the No. 1 Squadron and No. 6 Squadron of the No. 82 Wing.

A8-125 USAF Serial Number 67-0125 (D1-01)
After its first flight on August 28, 1968, the aircraft was initially delivered on September 4, 1968. It was accepted by the USAF (DD-250 date) on August 28, 1969. After storage at General Dynamics Fort Worth, the aircraft arrived at Amberley on June 1, 1973. (Author's Collection)

A8-126 USAF Serial Number 67-0126 (D1-02)
Above: The aircraft first flew on July 13, 1968. This was the first flight of an F-111C. It was accepted by the USAF (DD-250 date) on September 6, 1968. After storage at General Dynamics, Fort Worth, the aircraft arrived at Amberley on June 1, 1973. A8-126 was the first F-111C modified to an RF-111C. It was delivered to General Dynamics for modification in October 1978. The conversion was completed in May 1979, arriving back at RAAF Amberley in August 1979. It is seen here at Melbourne on February 2, 1985. (APA-AVN Photo Australia)

A8-127 USAF Serial Number 67-0127 (D1-03)
Right: The aircraft first flew on July 17, 1968, and was delivered on August 28, 1969. It was accepted by the USAF (DD-250 date) on August 30, 1969. After storage at General Dynamics, Fort Worth, the aircraft arrived at Amberley on June 1, 1973. It crashed on September 13, 1993, at Black Mountain near Guyra, New South Wales during night simulated attack. The aircraft hit the ground during an escape maneuver, killing both crew members, Flight Lieutenant Jeremy McNess and Flight Lieutenant Mark Cairns-Cowan. (Author's Collection)

CHAPTER SIX: ROYAL AUSTRALIAN AIR FORCE F-111s

A8-128 USAF Serial Number 67-0128 (D1-04)
The aircraft first flew on July 30, 1968. It was accepted by the USAF (DD-250 date) on August 30, 1969. After storage at General Dynamics, Fort Worth, the aircraft arrived at Amberley on June 1, 1973. It is seen here in flight with the light gray underside worn by all early F-111s. It crashed April 2, 1987, near Tenterfield, New South Wales during night simulated attack. Both crew members, Flight Lieutenant Mark Fallon and Flying Officer William Pike, were killed. (Marty Isham's Collection)

A8-129 USAF Serial Number 67-0129 (D1-05)
The aircraft first flew on August 5, 1968. It was accepted by the USAF (DD-250 date) on August 30, 1969. After storage at General Dynamics, Fort Worth, the aircraft arrived at Amberley on June 1, 1973. (Lenn Bayliss)

A8-130 USAF Serial Number 67-0130 (D1-06)
The aircraft first flew on September 13, 1968. It was accepted by the USAF (DD-250 date) on September 3, 1969. After storage at General Dynamics, Fort Worth, the aircraft arrived at Amberley on June 1, 1973. (Author's Collection)

A8-131 USAF Serial Number 67-0131 (D1-07)
The aircraft first flew on October 22, 1968. It was accepted by the USAF (DD-250 date) on September 16, 1969. After storage at General Dynamics, Fort Worth, the aircraft arrived at Amberley on July 12, 1973. (Author's Collection)

A8-132 USAF Serial Number 67-0132 (D1-08)
The aircraft first flew on October 21, 1968. It was accepted by the USAF (DD-250 date) on September 25, 1969. After storage at General Dynamics, Fort Worth, the aircraft arrived at Amberley on July 12, 1973. A8-132 is instrumented for flight test purposes and is used to test various aircraft modifications and for integration of new weapons. It returned to Amberley (June 9, 1996) from McClellan AFB CA after undergoing Avionics Upgrade Program (AUP) flight test. The aircraft successfully fired Harpoon missile at Barking Sands Test Facility, Hawaii (June 1996). During the testing, as seen in this photo taken at Edinburgh, Australia on May 30, 1985, the sides were painted light gray in color to ease photography of weapons releases. The aircraft was also the P109 engine change test aircraft. (APA-AVN Photo Australia)

A8-133 USAF Serial Number 67-0133 (D1-09) NO PHOTO AVAILABLE
The aircraft first flew on October 22, 1968. It was accepted by the USAF (DD-250 date) on September 22, 1969. After storage at General Dynamics, Fort Worth, the aircraft arrived at Amberley on July 12, 1973. A8-133 was destroyed September 29, 1977 after birdstrike (three pelicans) through the windshield at Evans Head Weapons Range New South Wales. The ejection was out of parameters; both crew members, Squadron Leader John Holt and Flight Lieutenant A.P. "Phil" Noordink, were killed.

A8-134 USAF Serial Number 67-0134 (D1-10)
Right: The aircraft first flew on November 18, 1968. It was accepted by the USAF (DD-250 date) on September 22, 1969. After storage at General Dynamics, Fort Worth, the aircraft arrived at Amberley on July 12, 1973. It was modified by the 3rd Aircraft Depot at Amberley to RF-111C configuration during 1980. (Lenn Bayliss)

A8-135 USAF Serial Number 67-0135 (D1-11)
The aircraft first flew on December 2, 1968. It was accepted by the USAF (DD-250 date) on September 25, 1969. After storage at General Dynamics, Fort Worth, the aircraft arrived at Amberley on July 12, 1973. (82nd Wing Photographic, RAAF)

A8-136 USAF Serial Number 67-0136 (D1-12) NO PHOTO AVAILABLE
The aircraft first flew on December 5, 1968. It was accepted by the USAF (DD-250 date) on September 22, 1969. After storage at General Dynamics, Fort Worth, the aircraft arrived at Amberley on July 12, 1973. It crashed on April 27, 1977, near Armidale, New South Wales, after an engine fire. Both crew members survived the ejection.

CHAPTER SIX: ROYAL AUSTRALIAN AIR FORCE F-111s

A8-137 USAF Serial Number 67-0137 (D1-13)
The aircraft first flew on December 12, 1968. It was accepted by the USAF (DD-250 date) on September 22, 1969. After storage at General Dynamics, Fort Worth, the aircraft arrived at Amberley on October 1, 1973. It is seen here at McClellan AFB in September 1973, just prior to delivery to the RAAF. A8-137 crashed August 24, 1979, at Ohakea, New Zealand after double engine failure on takeoff. Both crew members survived the ejection. (Author)

A8-138 USAF Serial Number 67-0138 (D1-14)
The aircraft first flew on December 17, 1968. It was accepted by the USAF (DD-250 date) on September 22, 1969. After storage at General Dynamics, Fort Worth, the aircraft arrived at Amberley on October 1, 1973. (Lenn Bayliss)

A8-139 USAF Serial Number 67-0139 (D1-15) NO PHOTO AVAILABLE
The aircraft first flew on December 18, 1968. It was accepted by the USAF (DD-250 date) on September 25, 1969. After storage at General Dynamics, Fort Worth, the aircraft arrived at Amberley on October 1, 1973. It crashed into the sea near Moruya, New South Wales, on January 28, 1986, at night during simulated attack. Both crew members, pilot Flight Lieutenant Stephen Erskine RAAF and WSO Captain Gregory S. Angell USAF, were killed. The WSO was a USAF exchange officer.

A8-140 USAF Serial Number 67-0140 (D1-16)
The aircraft first flew on December 24, 1968. It was accepted by the USAF (DD-250 date) on September 30, 1969. After storage at General Dynamics, Fort Worth, the aircraft arrived at Amberley on October 1, 1973. (Lenn Bayliss)

A8-141 USAF Serial Number 67-0141 (D1-17) NO PHOTO AVAILABLE
The aircraft first flew on December 18, 1968. It was accepted by the USAF (DD-250 date) on September 30, 1969. After storage at General Dynamics, Fort Worth, the aircraft arrived at Amberley on October 1, 1973. The aircraft crashed in Auckland Harbor, New Zealand on October 25, 1978, after a fire. Both crew members survived the ejection.

A8-142 USAF Serial Number 67-0142 (D1-18)
Right: The aircraft first flew on December 18, 1968. It was accepted by the USAF (DD-250 date) on September 30, 1969. After storage at General Dynamics, Fort Worth, the aircraft arrived at Amberley on October 1, 1973. The aircraft was the first production AUP aircraft which test flew at Amberley on February 27, 1996. It is seen here at Nellis AFB in May 1997, participating in the Golden Air Tattoo for the USAF 50th Anniversary. It is carrying an AGM-142 Have Nap (Popeye) missile. (Author)

A8-143 USAF Serial Number 67-0143 (D1-19)
The aircraft first flew on January 7, 1969. It was accepted by the USAF (DD-250 date) on October 7, 1969. After storage at General Dynamics, Fort Worth, the aircraft arrived at Amberley on December 4, 1973. It was modified by the 3rd Aircraft Depot at Amberley to RF-111C configuration during 1980. It's seen here at Bergstrom AFB, Texas at RAM 86. (George Cockle)

Seen in 1997 at Amberley in the markings on No 1 Squadron. (Lenn Bayliss)

CHAPTER SIX: ROYAL AUSTRALIAN AIR FORCE F-111s

A8-144 USAF Serial Number 67-0144 (D1-20)
The aircraft first flew on January 2, 1969. It was accepted by the USAF (DD-250 date) on November 4, 1969. After storage at General Dynamics, Fort Worth, the aircraft arrived at Amberley on December 4, 1973. (Author's Collection)

A8-145 USAF Serial Number 67-0145 (D1-21)
The aircraft first flew on December 31, 1968. It was accepted by the USAF (DD-250 date) on November 4, 1969. After storage at General Dynamics, Fort Worth, the aircraft arrived at Amberley on December 4, 1973. (Lenn Bayliss)

A8-146 USAF Serial Number 67-0146 (D1-22)
The aircraft first flew on January 2, 1969. It was accepted by the USAF (DD-250 date) on October 14, 1969. After storage at General Dynamics, Fort Worth, the aircraft arrived at Amberley on December 4, 1973. During 1980, the aircraft was modified by the 3rd Aircraft Depot at Amberley to RF-111C configuration. It is seen here in 1993, in markings celebrating No. 6 Squadron's 20 years of F-111 operations (Lenn Bayliss)

A8-147 USAF Serial Number 67-0147 (D1-23)
The aircraft first flew on January 9, 1969. It was accepted by the USAF (DD-250 date) on November 18, 1969. After storage at General Dynamics, Fort Worth, the aircraft arrived at Amberley on December 4, 1973. (APA-AVN Photo Australia)

A8-148 USAF Serial Number 67-0148 (D1-24)
The aircraft first flew on January 17, 1969. It was accepted by the USAF (DD-250 date) on November 18, 1969. After storage at General Dynamics, Fort Worth, the aircraft arrived at Amberley on December 4, 1973. (Lenn Bayliss)

F-111A

A8-109 USAF 67-0109 (A1-154)
The aircraft arrived at Amberley on August 20, 1982, an ex-USAF aircraft, as an attrition replacement aircraft. It has patches from damage caused by North Vietnamese ground fire damage received during Linebacker II. It was modified by No. 3 Aircraft Depot, RAAF Amberley to F-111C standard. (82nd Wing Photographic Services, RAAF)

CHAPTER SIX: ROYAL AUSTRALIAN AIR FORCE F-111s

A8-112 USAF 67-0112 (A1-157)
The aircraft arrived at Amberley on January 15, 1983, an ex-USAF aircraft, as an attrition replacement aircraft. It was modified by No. 3 Aircraft Depot, RAAF Amberley to F-111C standard. (Lenn Bayliss)

A8-113 USAF 67-0113 (A1-158)
The aircraft arrived at Amberley on May 23, 1982, an ex-USAF aircraft, as an attrition replacement aircraft. It was modified by No. 3 Aircraft Depot, RAAF Amberley to F-111C standard. (Author's Collection)

A8-114 USAF 67-0114 (A1-159)
The aircraft arrived at Amberley on May 23, 1982, an ex-USAF aircraft, as an attrition replacement aircraft. It was modified by No. 3 Aircraft Depot, RAAF Amberley to F-111C standard. (Author's Collection)

63-9768 (A1-03) NO PHOTO AVAILABLE
A pre-production F-111A (3rd built), the aircraft arrived at Amberley in 1995 by sea transport to be used for ground training. It has ejection seats fitted instead of the crew module. The aircraft is on permanent loan to the RAAF.

F-111G

A8-259 USAF 68-0259 (B1-31) NO PHOTO AVAILABLE
The aircraft arrived at Amberley during October 1993.

A8-264 USAF 68-0264 (B1-36)
Right: The aircraft arrived at Amberley on February 11, 1994. It's seen here in storage (82nd Wing Photographic, RAAF)

A8-265 USAF 68-0265 (B1-37)
Below: The aircraft arrived at Amberley on September 24, 1993. It's seen here on departure from McClellan AFB on the way to Australia. (Joe Arnold)

A8-270 USAF 68-0270 (B1-42)
The aircraft was the first F-111G delivered to the RAAF, arriving at Amberley on September 24, 1993. It is seen here taxiing out for departure from McClellan AFB. (Joe Arnold)

CHAPTER SIX: ROYAL AUSTRALIAN AIR FORCE F-111s

A8-271 USAF 68-0271 (B1-43)
The aircraft arrived at Amberley on January 14, 1994, and placed in storage on May 4, 1994. (Dave Riddel)

A8-272 USAF 68-0272 (B1-44)
Nicknamed Boneyard Wrangler, the aircraft arrived at Amberley on May 10, 1994, being flown by Group Captain Dave Dunlop and Flight Lieutenant Dave Riddel from SM-ALC. It was the first RAAF F-111G to be recovered to flying condition from AMARC. As seen here, this aircraft has special tail art. (Dave Riddel)

A8-274 USAF 68-0274 (B1-46)
The aircraft arrived at Amberley on December 6, 1993. (Dave Riddel)

A8-277 USAF 68-0277 (B1-49) NO PHOTO AVAILABLE
The aircraft arrived at Amberley on March 25, 1994, and placed in storage on May 4,1994.

A8-278 USAF 68-0278 (B1-50)
The aircraft arrived at Amberley on May 10, 1994, and placed in storage. It's seen here in storage. (82nd Wing Photographic, RAAF)

A8-281 USAF 68-0281 (B1-53)
The aircraft arrived at Amberley on December 6, 1993. Painted in RAAF 75th Anniversary markings, A8-281 is seen here in June, 1996, at Amberley. (Lenn Bayliss)

A8-282 USAF 68-0282 (B1-54)
The aircraft arrived at Amberley on March 25, 1994. It's seen here in storage. (82nd Wing Photographic, RAAF)

CHAPTER SIX: ROYAL AUSTRALIAN AIR FORCE F-111s

A8-291 USAF 68-0291 (B1-63) NO PHOTO AVAILABLE
The aircraft arrived at Amberley on October 22, 1993. It was the first RAAF F-111 to complete R5 (PDM) at SM-ALC McClellan AFB in July 1996.

A8-506 USAF 69-6506 (B1-68) NO PHOTO AVAILABLE
It was stored at Sacramento ALC awaiting delivery to Australia.

A8-512 USAF 69-6512 (B1-74) (Lenn Bayliss)

A8-514 USAF 69-6514 (B1-76) (Lenn Bayliss)

CHAPTER SEVEN

British Royal Air Force F-111s

The F-111 production contract issued in May 1967, included 50 F-111s for the United Kingdom. These aircraft were designated F-111Ks. The British Government withdrew from the F-111 program and canceled their order in January 1968. At the time of the cancellation, only two aircraft had been started and were never completed. They were salvaged with their assemblies used on other F-111s. Basically, the F-111K was an F-111A with a higher gross weight takeoff capability (using the landing gear of the FB-111), fire control/navigation system updated with Mark II and FB-111 components, British furnished mission and traffic control systems, and a reconnaissance conversion capability. In addition, the F-111K had a fuselage centerline pylon capability and could carry four 600 gallon external fuel tanks.

Two configurations of F-111Ks were to be delivered, a Trainer/Strike configuration and a Strike/Recon configuration. The airframes of these two types were identical and differed only in crew module avionics and flight controls, weapons bay, and provisions for alternate loads. One Strike/Recon and one Trainer/Strike aircraft were to be built as the two Development, Testing, and Evaluation (DT&E) aircraft. F-111K #1 was a Strike/Recon configured aircraft to be used for airframe and weapons separation testing. F-111K #2 was to be configured as a Trainer/Strike version to be used for avionics and weapon separation testing. Both were to be refurbished and redelivered as production aircraft.

The nose of the F-111K was a MK II configuration, common to the F-111D and the FB-111A. The forward electronics bay was peculiar to the F-111K. It housed an inflight refueling probe oriented fore and aft on the upper centerline, hinged at the base of the windshield and could be extended upward and back. This differed from the F-111B refueling probe which laid sideways, hinged at the aircraft centerline in front of the pilot. When deployed, both probe types somewhat resembled the refueling probe of the U.S. Navy A-6. The nose section also had three camera windows on the underside forward of the nose landing gear. The weapons bay of the Strike/Recon version contained a removable centerline mounted pylon which was attached to the top of the bay and extended down through special cutouts in the bay doors. The pylon carried one British designed Ejector Release Unit (ERU) weapons ejector rack. A Reconnaissance Pallet which was British Furnished Equipment (BFE) could be installed in the weapons bay of the Strike/Recon version.

The following time line relates the history of the F-111K program and the history of the two airframes.

July 26, 1967
The first production F-111K entered fuselage mating. It was scheduled to go to Bay 4 of the special projects hangar for manufacturing completion.

August 1967
Fort Worth was advised that the RAF had elected to install the FB-111 modified APQ-128 (TFR) effective F-111K No. 3 and on.

September 18, 1967
F-111K No. 1 completed major mating and began final assembly.

October 27, 1967
F-111K No. 2 was mated and scheduled to go to final assembly.

December 1967
F-111K No. 2 completed factory major mating and began final assembly.

January 16, 1968
British Prime Minister Harold Wilson announced the cancellation of the UK order for 50 F-111Ks. As a result of the cancellation of the 50 F-111Ks by the United Kingdom, Fort Worth was directed to investigate the conversion of all Ks to FBs and conversion of about half of the Ks to additional As and the remainder to FBs.

CHAPTER SEVEN: BRITISH ROYAL AIR FORCE F-111s

(Lockheed Martin Tactical Aircraft Systems)

February 1968
Fort Worth was directed to divert F-111K No. 1 and No. 2 to the F-111A program as test aircraft and to designate them as YF/F-111A No. 1 and 2, increasing the total quantity of planned F-111As from 235 to 237. Notification was also received to cancel F-111Ks No. 3 through 50 and to divert the common assets to the FB-111A program.

May 1968
A plan was submitted to SPO to utilize YF-111A No. 1 as a high speed, flutter, vibration and acoustic test aircraft, and YF-111A No. 2 as a stability and flight control system test aircraft.

July 29, 1968
SPO and Air Force decided that it was not in their best interests to convert the two UK aircraft to any test or operational configuration. Accordingly, Fort Worth was directed to submit a cancellation cost proposal on the basis that F-111K (YF-111A) No. 1 and No, 2 would be salvaged (dispositioned as component parts, etc.).

November 1968
Fort Worth Division received from SPO contractual direction to salvage F-111K Nos. 1 & 2 and to utilize the left-hand wing from F-111K No. 1 for testing the F-111 flap and slat system in lieu of fabricating a dummy wing.

January 1969
Modification and instrumentation of the No. 1 F-111K's left wing was in process for use on the high-lift endurance test stand.

May through September 1969
F-111Ks Nos. 1 and 2 were de-mated and disassembled into major airframe components and subsystems for USAF applications. The airframe components and subsystems were to be further disassembled as required to obtain usable parts, spare items, GFAE (Government Furnished Assets & Equipment), and residual scrap. The major components were to be preserved and stored. By August, de-mating of both F-111K No. 1 and No 2 had been completed. F-111K No. 1 has been disassembled. No. 1's wing carry-through fittings would be used for engineering tests, with No. 2's fittings used for production. By September, de-mating and disassembly of both F-111Ks was complete. General Dynamics retained the crew module from F 111K No. 1 and saved the forward center fuselage from F-111K No, 2 for possible use in the construction of an RF-111 Engineering Tooling Manufacturing Aid (ETMA). The crew module from F-111K No. 2 was authorized for possible use as a trainer, for mock-up or engineering test purposes. Wing carry-through box from F-111K No. 2 was reworked and used in F-111E 68-0073.

February 1970
A determination was made that all of the F-111K No. 1 and No. 2 major subsystem items, except the canopies and transparencies, would be turned over to Sacramento Air Material Area (SMAMA).

March 1970
The gear (main and nose) from F-111K No. 2 were used for replacement parts on the refused takeoff tests. Component parts from the main gear on F-111K No. 1 were provided to General Dynamics Engineering for replacement of parts damaged during the heavy gear qualification and fatigue tests.
April 1970. All F-111K assets had been disposed of with the exception of one main landing gear disassembled for components required to continue the main landing gear (heavy) fatigue tests.

67-0149/XV884 E1-01
The aircraft was to be a Strike/Recon version (F-111K) and was not completed at the time of the contract cancellation. It was later disassembled. The crew module was used as part of the RF-111 program.

67-0150/XV885 E1-02

The aircraft was to be a Trainer/Strike version (TF-111K) and was not completed. It was disassembled after the contract was canceled. The wing box was used in F-111E 68-0083, the forward and center fuselage was used as part of the RF-111 program, and the crew module was used as a trainer.

TF-111K Canceled/Construction not started.
67-0151 to 67-0152/XV886 to XV887

F-111K Canceled/Construction not started.
68-0152 to 68-0158/XV902 to XV947
68-0181 to 68-0210
68-0229 to 68-0238

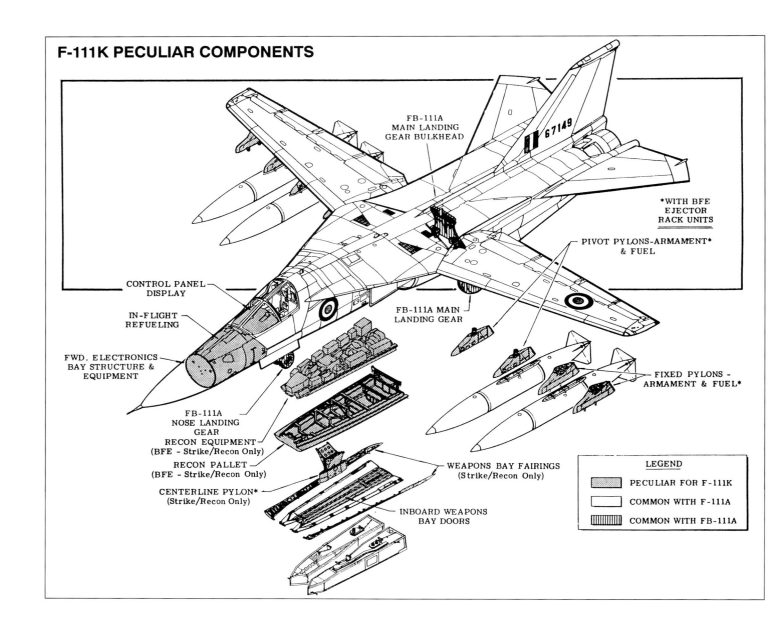

CHAPTER SEVEN: BRITISH ROYAL AIR FORCE F-111s

These two photos show the progress of the two F-111Ks at contract cancellation on March 20, 1968. (Lockheed Martin Tactical Aircraft Systems)

CHAPTER EIGHT

Combat Operations

Combat Lancer (March 15, 1968 to November 22, 1968)

The Combat Bullseye I tests of early 1967, clinched the USAF decision to rush a small detachment of seven F-111As to Southeast Asia, under the name Combat Lancer, to boost night and all-weather attacks while testing the aircraft's overall combat capability. Combat Lancer was preceded by Harvest Reaper, started in June 1967, to correct known F-111A shortcomings and prepare the aircraft for combat. The Harvest Reaper modifications (mainly more avionics and electronic countermeasures (ECM) equipment) would enter the F-111A production lines, if successfully proven in combat. Combat Lancer looked to another pre-combat project (Combat Trident) for trained pilots. Combat Trident training of the aircrews completed on March 6th, only nine days before the Combat Lancer deployment on March 15, 1968. The six aircraft deployed were 66-0016, 66-0017, 66-0018, 66-0019, 66-0021, and 66-0022. They reached Takhli Royal Thai Air Base on the 17th. At month's end, after 55 missions that centered on North Vietnam targets, two aircraft 66-0017 and 66-0022 had been lost. Two more aircraft 66-0024 and 66-0025 arrived as replacements. The loss of a third Combat Lancer aircraft on April 22 halted F-111A operations. The surviving five aircraft finally returned to Nellis on November 22.

66-0016 flew the first Combat Lancer mission on March 18, 1968, attacking a truck park and storage area with 12 Mk 117 750 pound bombs. 66-0017 flew its first Combat Lancer mission on March 27. It crashed and was destroyed on March 30, 1968, during Combat Lancer mission. Its callsign was Hotrod 76. The crew, Major Sandy Marquardt and Capt Joe Hodges, successfully ejected and was recovered. The cause of the aircraft loss was a tube of sealant lodged in the horizontal stabilizer actuator. 66-0018's first mission was on March 25, 1968. 66-0019, on March 27, during Combat Lancer, had a hydraulics problem which gave it the distinction of being the first F-111 combat ground abort. The aircraft arrived at Tahkli on March 27. 66-0022 was the first combat related loss, callsign Omaha 77, destroyed March 28, 1968, three days after Combat Lancer operations started. The target was the Chao Hao truck park. Killed were Major Henry McCann and Captain Dennis Graham.

66-0024, a Combat Lancer replacement aircraft, deployed to Takhli on April 1, 1968. Flying as callsign Tailbone 78, aircraft 66-024 crashed and was destroyed April 22, 1968, becoming the third and last jet lost during Combat Lancer. Neither the aircraft, nor crew, (Commander David "Spade" Cooley USN, and Lt. Col. Ed Palmgren), was found. 66-0025 a Combat Lancer replacement aircraft was deployed to Takhli for Combat Lancer on April 1, 1968.

F-111F 74-0187 was marked with combat streamers of the F-111's combat operations for the naming and farewell ceremony. It also was marked with the tail codes of the tactical wings (HG, LN, MO, NA, UH, and WA) and Wing numbers of the SAC Bombardment Wings (380th BW and 509th BW). Though missing from the nose CC of the 27th Fighter Wing was carried on the tail. (Lockheed Martin Tactical Aircraft Systems)

The aircraft taking part in Combat Lancer all carried the special rudder markings seen here. (Lockheed Martin Tactical Aircraft Systems)

This Combat Lancer F-111A is seen in its revetment at Takhli RTAFB loaded with 24 MK 82 500 pound bombs (Author's Collection)

Constant Guard/Linebacker (September 27, 1972 to Mid-June 1972)

When the North Vietnamese invaded South Vietnam in the spring of 1972, the 474th TFW began seeing indications of a possible deployment, since their F-111As were the only combat-ready F-111s at the time. It was decided that when the deployment occurred, a record for the elapsed time between the alert order and the bombs on target would be attempted. The 429th TFS was the first squadron to deploy. They divided into three groups to deploy. The first two groups acted as ferry crews moving the aircraft from Nellis to Takhli Royal Thai Air Force Base, Thailand. The third group flew directly to Takhli and prepared to fly the first combat missions. The combat crews had about 24 hours of rest before the airplanes arrived. The 12 aircraft deployed on September 27, 1972. The record was set with F-111As in combat 55 miles northwest of Hanoi – 33 hours after leaving Nellis. However, one of the aircraft, (67-0078), was lost in the first night of operations. The aircraft made it to the target, but did not return home. The first mission package was scheduled for six aircraft and assigned targets in the western part of North Vietnam. The aircraft had been on the ground about four hours before launching on their bombing missions. Three of the six aircraft ground-aborted the mission with equipment failure; the fourth aircraft aborted in the air after ECM equipment failed; the fifth airplane(67-0078) never returned from the mission; and the sixth aircraft couldn't get to its primary target and was forced to bomb an alternate.

The deployment occurred in three phases. The first phase consisted of twelve 429th TFS F-111As deploying from Nellis on September 27, 1972. The following F-111As deployed to Tahkli RTAFB in the first phase: 67-0060, 67-0065, 67-0068, 67-0070, 67-0072, 67-0074, 67-0075, 67-0078, 67-0079, 67-0083, 67-0084, and 67-0086. These aircraft arrived at Tahkli on October 1, 1972.

The second phase was an additional twelve 429th TFS aircraft. The following F-111As also left Nellis AFB for Tahkli on September 27: 67-0059, 67-0061, 67-0063, 67-0064, 67-0066, 67-0069, 67-0071, 67-0076, 67-0080, 67-0087, 67-0088, and 67-0090.

67-0098, one of the-111A deployed for Constant Guard, was photographed in its revetment at Takhli RTAFB. (Harley Copic)

The third phase consisted of twenty-four 430th TFS aircraft. The following F-111As left Nellis on September 29, 1972: 67-0073, 67-0077, 67-0085, 67-0091, 67-0092, 67-0093, 67-0094, 67-0095, 67-0096, 67-0097, 67-0098, 67-0099, 67-0100, 67-0101, 67-0102, 67-0103, 67-0104, 67-0105, 67-0106, 67-0107, 67-0110, 67-0112, 67-0113, and 67-0114. The aircraft arrived at Tahkli on October 4, 1972. 67-0092 and 67-0110 had minor problems enroute and arrived on October 5, 1972.

Replacement aircraft 67-0062 and 67-0067 arrived on October 23, 67-0058 arrived on November 21, 67-0089 arrived on November 25, 67-0109 arrived on December 11, and 67-0081 and 67-0111 arrived on May 8 of the following year.

After the loss on the first night, the deployed units stood down for five days, while the aircraft systems were checked out. The first combat missions resumed on October 5, 1972, were medium altitude drops on the passes just over the North Vietnamese border. By mid October the F-111s were once again attacking targets in high threat areas. The 111s flew at night and solo. Four F-111As could deliver the bomb loads of 20 F-4s. They flew without Wild Weasel or EB-66 electronic countermeasure escort aircraft or KC-135 tankers (needed by the F-4s which they replaced) and in all weather. F-111As flew 20 strikes over North Vietnam on November 8, in weather that grounded other aircraft. In addition to dropping their own bombs, the F-111s also served as pathfinders, directing other fighters where to drop their bombs. They also participated in the Linebacker II bombing campaign between December 18 and 29, 1972. During Constant Guard V/Linebacker (September 28, 1972 through the end of March 1973) the 18 F-111As from the 429th and 430th flew over 4,0000 combat sorties with only six combat losses. The resulting 0.015 per cent loss rate made the F-111A the most survivable combat aircraft of the Vietnam War.

68-0078 was lost on the first night of bombing raids – September 28; 67-0066 was lost on October 17; 67-0063 on November 8; 67-0092 on November 21; 67-0099 on December 18; and 67-0068 on December 27.

67-0071 and 67-0098 were involved in a mid-air collision on February 17, 1973, while changing lead and acting as pathfinders for a flight of F-4s. There were also two operational non-combat losses (67-0072 on February 20 and 67-0111 on June 16, 1973). The 428th TFS replaced the 430th on January 2, 1973.

The air war in Vietnam ended following the January 1973 peace agreement with North Vietnam. By the end of July 1974, the F-111As had moved from Takhli to Korat RTAFB and with the 429th TFS and 428th TFS transferring to the 347th TFW.

During the post Vietnam war era, the F-111As of the 347th TFW at Korat continued flying combat missions in Southeast Asia (mostly Cambodia and southern Laos). These missions were against communist forces trying to overthrow the Gov-

CHAPTER EIGHT: COMBAT OPERATIONS

The deployed F-111A wore HG tail codes while assigned to the 347th TFW at Korat RTAFB.

ernment of Cambodia. In mid-May 1975, F-111As located the SS Mayaguez after it was hijacked by Cambodian communists, and on May 14, the F-111s sank one of the gunboats escorting the SS Mayaguez. The two squadrons (428th and 429th TFS) remained at Korat until returning to the US in mid-June 1975.

El Dorado Canyon (April 14 and 15, 1986)

Early in 1986, F-111Fs of the 48th TFW began planning for possible counter-terrorist bombing attacks. In response to the simultaneous machine-gun killings of airline passengers in Rome and Vienna on December 17, 1985, the planning was refined. The operation was waiting for another terrorist incident with an obvious link to the prime suspect, Colonel Muammar Khadaffi's Libya. As a result of the bombing of the La Belle Disco in Berlin on April 5, 1986, the U.S. planned to go ahead. There were two basic plans: six aircraft, flying in international airspace, would attack; or, if overflight permission could be obtained from France, eighteen attacking aircraft would be used. The plan was changed with less than two days to go. Eighteen aircraft (plus six spares, which returned after the first inflight refueling) would attack through international airspace (around Spain) using massive inflight refueling support. A total of nineteen KC-10s and ten KC-135s launched from Mildenhall and Fairford to provide the seven million pounds of jet fuel required for this mission. This response reflected U.S. and British, from whose soil the raid was launched, determination to teach the sponsors of terrorism a lesson.

Below: This photo was taken at Lakenheath as the El Dorado Canyon aircraft lined up for takeoff. (Author's Collection)

The following F-111Fs took part in Operation El Dorado Canyon:

70-2392 Puffy 11	72-1449 Lujac 21	70-2390 Remit 31
70-2416 Puffy 12	71-0888 Lujac 22	72-1445 Remit 32
70-2394 Puffy 13	70-2387 Lujac 23	74-00178 Remit 33
73-00707 Puffy 14	70-2405 Lujac 24	70-2382 Remit 34
70-2403 Elton 41	70-2413 Karma 51	70-2371 Jewel 61
70-2396 Elton 42	70-2389 Karma 52	70-2383 Jewel 62
70-2363 Elton 43	71-0889 Karma 53	74-00177 Jewel 63
70-2404 Elton 44	70-2415 Karma 54	70-2386 Jewel 64

Operation El Dorado Canyon was launched on the night of April 14, 1986. In addition to the twenty-four F-111Fs, six EF-111As of the 42nd ECS would be used. Four aircraft accompanied the strike flights. One spare remained with the tankers. Six aircraft were prepared for the mission, including a ground spare. The six aircraft were: 66-0030, 66-0033, 66-0057, 67-0034, 67-0041, and 67-0052.

Nine aircraft attacked the Azziziyah barracks in Tripoli (where Khadaffi lived); Six attacked Tripoli's airport; while three others attacked the terrorist training camp at Sidi Bilal, west of Tripoli. Originally, the plan had been for six aircraft to go against each target, but this was changed at the last minute by the White House.

The plan, as executed, had nine aircraft attacking the barracks in single file. The defenders had been alerted and apparently were able to shoot down the eighth aircraft (70-2389) – the crew was killed. Even though one aircraft was lost, the raid was the most successful military action by the United States since the Vietnam War. Also, it convinced the hostile third-world political leaders that they could no longer engage in terrorism with impunity.

Desert Shield/Desert Storm (August 1990 to February 27th 1991)

On August 2, 1990, the United States began Operation Desert Shield to protect Saudi Arabia following the Iraqi invasion of Kuwait. The first USAFE Wing to deploy from Europe was the 48th TFW from RAF Lakenheath. The initial contingent of 19 aircraft left for Taif on August 25, 1980. It was made up contingents of both the 492nd TFS and the 494th TFS. A second group of 14 followed on September 25, made up mostly of aircraft of the 493rd TFS. The remainder of the 492nd TFS and 494th TFS, which stayed behind at Lakenheath, was reorganized as the 492nd TFS. This new 492nd TFS was deployed with 12 aircraft to Taif on November 29. The next group of 11 aircraft deployed on December 11. A final group of six aircraft transferred to Taif brought the total to 67 aircraft deployed at the start of Operation Desert Storm on January 17, 1991.

The following aircraft were in place at Taif on January 17, 1991:

70-2362, 70-2363, 70-2364, 70-2365, 70-2369, 70-2370, 70-2371, 70-2378, 70-2379, 70-2383, 70-2384, 70-2386, 70-2387, 70-2390, 70-2391, 70-2392, 70-2394, 70-2396, 70-2398, 70-2399, 70-2401, 70-2402, 70-2403, 70-2404, 70-2405, 70-2406, 70-2408, 70-2409, 70-2411, 70-2412, 70-2413, 70-2414, 70-2415, 70-2416, 70-2417, 70-2419, 71-0883, 71-0884, 71-0885, 71-0886, 71-0887, 71-0888,

(Craig Brown)

CHAPTER EIGHT: COMBAT OPERATIONS

(Craig Brown)

71-0889, 71-0890, 71-0891, 71-0892, 72-1442, 72-1443, 72-1444, 72-1445, 72-1446, 72-1448, 72-1449, 72-1450, 72-1451, 72-1452, 73-0708, 73-0710, 73-0712, 73-0715, 74-0177, 74-0178, 74-0180, 74-0181, 74-0182, 74-0184, 74-0185.

70-2384 had originally deployed, but was damaged in a collision with a KC-135 on the first night of the air war and was replaced by 70-2369. 74-0183 had deployed to Taif, but crashed prior to Desert Storm, on October 10, 1990, during training at Askr Bombing Range near Taif. The crew was killed in the crash.

The 48th TFW (Provisional) was established to control the aircraft at Taif. Aircraft and personnel were formed into units at Taif and adopted the following names: 492nd Justice, 493rd Freedom, 494th Liberty, and 495th Independence.

During Desert Storm, the F-111Fs dropped nearly 80 percent of the war's Laser Guided Bombs (LGBs). The 48th TFW(P) flew 2,417 sorties for 9,381.2 hours. They dropped 469 GBU-10, 389 GBU-101, 2,542 GBU-12, 270 GBU-24, 924 GBU-24A and two GBU-28 laser-guided bombs; 62 Imaging Infrared (IIR) and eight Electro-Optical (EO) guided GBU-15 bombs; 530 CBU-87, 212 CBU-89 cluster bombs, and 12 Mk 82 and 146 Mk 84 general purpose bombs. Some of the of GBU-15s were used to stop the flow of oil from a Kuwaiti oil refinery which was sabotaged by the Iraqis. On one of the oil refinery missions, being flown by two F-111Fs, the WSO in one aircraft guided GBU-15 from both aircraft to the target after his wingman's data link pod used to send steering commands to the GBU-15 failed. 245 of the 375 aircraft shelters (and the 141 aircraft in them) destroyed during the war were destroyed by the F-111Fs. The GBU-12 LGBs were used on anti-armor missions. The first two GBU-28 Deep Throat bunker-buster bombs to be used in combat were dropped by F-111F 70-2391 on the final night of the war.

Eighteen EF-111As were also controlled by the 48th TFW(P). Twelve of these Ravens (66-0014, 66-0027, 66-0030, 66-0033, 66-0037, 66-0044, 66-0046, 66-0057, 67-0037, 67-0038, 67-0039, and 67-0041) were deployed from the 390th ECS at Mountain Home AFB in August 1990. Five, (66-0016, 66-0023, 66-0038, 66-0050, and 66-0056), deployed from the 42nd ECS at Upper Heyford which transferred to the 48th TFW(P) on the first day of the war. The EF-111As flew 875 missions for 4,401.4 flight hours. An EF-111A from the 390th ECS (66-0016), flown by pilot Captain Jim Denton with EWO Captain Brent Brandon, scored an unofficial kill on the first night of the war when they maneuvered a pursuing Mirage F1EQ into the ground. The only EF-111 lost (66-0023 from the 42nd ECS) crashed on February 14. The crew was killed.

The 48th TFW(P) was inactivated during March 1991, with command of the elements of the 492nd TFS remaining at Taif until May 10, 1991, reverting to the 48th TFW at Lakenheath.

Operation Proven Force

F-111E units from the 20th TFW deployed to Incirlik AB, Turkey as part of Operation Proven Force, the code name by which operations from Turkey during the Gulf War were known. These F-111E operations were administered by the 7440th Wing (Provisional) which had been established at Incirlik. The Wing, equipped mostly with USAFE aircraft had twenty-two F-111Es assigned: 67-0120, 67-0121, 68-0004,

(USAF)

68-0005, 68-0013, 68-0015, 68-0016, 68-0017, 68-0026, 68-0029, 68-0031, 68-0039, 68-0040, 68-0046, 68-0049, 68-0050, 68-0061, 68-0068, 68-0069, 68-0072, 68-0074 and 68-0076. The aircraft were drawn from all the squadrons of the 20th TFW at Upper Heyford, and with most of the initial aircrews coming from the 79th TFS "Tigers." Aircrews from the 55th TFS and 77th TFS arrived later. Both AMP and non AMP F-111Es took part. Five EF-111As (66-0047, 66-0055, 67-0034, 67-0041, and 67-0042) from the 42nd ECS also participated in Proven Force.

Operation Provide Comfort

Operation Provide Comfort, intended to defend the Kurdish population in northern Iraq, began on April 5, 1991. It was supported by a 48th TFW aircraft and crews rotating from Lakenheath to Incirlik from September 25, 1991, until October 2, 1992. Provide Comfort operations were taken over by the 27th TFW at Cannon AFB, as it absorbed Lakenheath's F-111Fs. The EF-111As continued to fly operational missions in support of Operation Provide Comfort through 1996.

CHAPTER NINE

NASA F-111s

NASA's Dryden Flight Research Center (DFRC) at Edwards AFB flew four different F-111s (F-111A 63-9771, F-111A 63-9777, F-111A 63-9778, and F-111E 67-0115). They were part of an USAF/NASA program run by the DFRC.

F-111 Developmental Testing

The FRC's F-111A program was the only program of the 1960s that closely followed the earlier pattern of using NACA-NASA flight-test specialists to find and correct technical problems with a major new weapon system. The early F-111As had major engine problems, compressor surges and stalls. In January 1967, the Air Force sent the sixth production F-111A (63-9771) to FRC for testing. NASA pilots and engineers flew the airplane in an attempt to solve its problems. They studied the engine inlet dynamics of the aircraft to determine the nature of inlet pressure fluctuations that led to compressor surges and stalls. As a result of NASA, USAF, and General Dynamics testing, the engine problems were solved by a major inlet redesign, resulting in the Triple Plow 1 and Triple Plow II inlets. FRC's work had been crucial to this effort. The DFRC's second F-111A (63-9777), the twelfth built, arrived in April 1969, and was flown in a handling-qualities investigation program. Both aircraft were retired to MASDC at Davis Monthan in 1971.

Integrated Propulsion Control System (IPCS)

Another of the DFRC's test programs involving the F-111 was the Integrated Propulsion Control System (IPCS), using a USAF F-111E (67-0115). This program ran from March 1973, through February 1976. The aim of the IPCS was to accomplish for the propulsion system of an aircraft what fly-by-wire controls did for flight controls. Many factors affect engine performance. including throttle position, inlet position for variable-geometry inlets, fuel flow rates, and even the aircraft flight maneuvers. As with mechanical aerodynamic controls, the hydro-mechanical controls used in engine operation grew increasingly complex.

The Air Force Flight Propulsion Laboratory at Wright-Patterson AFB funded an experimental effort using a suitable airplane. A twin-engine airplane was required so it could be configured with one engine using normal controls and the other electronic controls. The engine which remained hydro-mechanically controlled for flight safety could also provide a comparison with the test engine. The F-111 being large, with two-seats, twin fanjet engines, a variable position inlet, and afterburners, was a natural for this test. In addition, it had an internal weapons bay that could be used to house the necessary electronic controls. The Air Force had an F-111 available, the first prototype of the General Dynamics F-111E series. Boeing was selected as prime contractor to develop the system, with Honeywell and Pratt & Whitney as subcontractors. NASA awarded the contracts for the Integrated Propulsion Control System program in March 1973.

The Flight Research Center received the F-111E in mid-1974, and began a series of 13 flights before modification to acquire a baseline data for comparison with results of the later IPCS tests. Installation of IPCS began in March 1975. The IPCS consisted of an instrumentation package, power supply, digital computer, and interface equipment installed in the weapons bay. The hydro-mechanical inlet and afterburner controls were replaced by new electronic controls.

The IPCS controlled only the F-111E's left engine. Hydro-mechanical control was available on the left engine for emergency backup use, and the right engine retained its own hydro-mechanical system. As a precaution, however, in the event of failure of the manually controlled engine during takeoff and the possibility of simultaneous problems with the experimental IPCS, all takeoffs were made toward Rogers Dry Lake, where an emergency landing could be made.

The F-111E completed its first IPCS flight on 4 September 1975, piloted by NASA's Gary Krier and the USAF's Stan Boyd. It completed an additional 14 IPCS investigations be-

Both 63-9771 (above) and 63-9777 (below) were used by NASA in F-111 Developmental Flight Testing. (Mick Roth)

fore the program concluded, making its last IPCS flight on February 27, 1976. NASA returned the F-111E to the USAF; restored to its original non-IPCS configuration, and later it served as a chase aircraft for the B-1 strategic bomber.

Transonic Aircraft Technology (TACT) and Advanced Fighter Technology Integration (AFTI) Programs

NASA and USAF interest in supercritical wing technology spawned the Transonic Aircraft Technology (TACT) program. An F-111A (63-9778) was modified to explore Super Critical Wing (SCW) technology and how it could benefit new military aircraft designs. During the 1960s, Langley Research Center had undertaken a great deal of wind-tunnel work on the F-111 aircraft. Also the F-111 was chosen as the testbed because of its variable-sweep wings. The new wings could be installed easily on the aircraft with a minimum of other modifications. Separate from this program, General Dynamics engineers had also conceived a retrofit program for the entire F-111 fleet. The company dubbed this program F-111 TIP (Transonic Improvement Program). By mid-1970, General Dynamics approached the USAF with this project. The

CHAPTER NINE: NASA F-111s

67-0115 was used by NASA in Integrated Propulsion Control System (IPCS) testing. (Dennis Jenkins)

Air Force wanted the F-111 tests as proof of concept evaluation, but did not want to retrofit the entire fleet. By mid-1971, NASA and General Dynamics had expended over 1600 hours of wind-tunnel test time on a suitable wing for the F-111. On June 16, 1971, NASA and the Air Force signed a joint Transonic Aircraft Technology (TACT) agreement to explore the application of supercritical wing technology to maneuverable military aircraft. The F-111 was flown at NASA's Flight Research Center, with development of the advanced configuration of the wing undertaken by NASA's Ames Research Center.

The F-111 was an ideal carrier for a Super-Critical Wing (SCW). Capable of supersonic speeds above Mach 2, the aircraft had a large volume for fuel and instrumentation. The wings were easily removable. The variable-sweep provision enabled SCW testing over a wide range of wing sweep angles and aspect ratios. Also, the Air Force planned to install pylons under the wings to carry external stores (such as bombs and drop tanks) to evaluate how these shapes interfered with the supercritical flow field. The 13th F-111A (63-9778) was available. NASA signed a loan agreement for the airplane with the Air Force on February 3, 1972, and on February 18, NASA flew the aircraft for the first time. The modified aircraft was ready for testing by the fall of 1973. On November 1, 1973, NASA made the first TACT flight, reaching Mach 0.85 at 8600 meters. On the 6th flight, March 20, 1974, they exceeded Mach 1; and on the 12th flight, they reached Mach 2.

The TACT aircraft flew frequently with a mixed Air Force-NASA crew. The wing design had definitely improved the performance of the F-111. At transonic speeds, the wing delayed drag rise and produced twice as much lift as the conventional F-111 wing. The supercritical wing did not impair high-Mach performance, with the aircraft flying much of the time above Mach 1.3. The external stores tests, with the F-111 carrying drag-inducing multiple bomb shapes on the pylons, were very successful countering fears that the external stores might wipe out any benefits from the supercritical planform.

After the initial TACT program, the F-111 TACT aircraft continued flying with a variety of aerodynamic experiments, including special shapes to evaluate base drag around the tail, experimental test instrumentation, and equipment destined for use with other airplanes. In 1980, seven years after

63-9778 was used in the Transonic Aircraft Technology (TACT) testing. As part of the program, the F-111s wings were replaced with new super critical wings. (NASA)

This top view of 63-9778 shows the revised shape of the TACT/AFTI wing. (NASA)

CHAPTER NINE: NASA F-111s

its first SCW exploration, it was still flying test missions. Because of the success of the TACT program, the Air Force Flight Dynamics Laboratory decided to proceed with another research effort called Advanced Fighter Technology Integration (AFTI). This was another joint Air Force-NASA effort, and consisted of various "Technology Sets." AFTI Tech Set II was a direct extension of the TACT program. Like TACT, the AFTI program involved the F-111. This TACT "second phase," later called AFTI F-111, went further, with concept of a "Mission Adaptive Wing (MAW)." This wing would not have the surface irregularities produced by conventional high-lift devices such as flaps and leading edge slats. Instead, internal mechanisms would flex and reshape the outer wing skin to produce a high-camber airfoil section for subsonic speeds, a supercritical section for transonic speeds, and a symmetrical section for supersonic speeds. This capability gave it the name of "mission adaptive." The TACT F-111 modified to the AFTI MAW demonstrator flew from Edwards AFB between 1984 and 1987.

As part of the AFTI program, 63-9778 was repainted in AFTI markings. (NASA)

63-97778 wears 13 on the nose gear doors, reflecting the fact that 63-9778 was the thirteenth F-111A built. (NASA)

Appendices

APPENDIX 1: INSIGNIA

APPENDICES

APPENDICES

APPENDIX 2: EXTERNAL DIFFERENCES

Model	Engines	Intake	Wings	Landing Gear
F-111A (Pre-Production)	YTF30-P-1 TF30-P-1 TF30-P-3	Translating Cowl and Splitter Plate	Short	Light Weight
F-111A	TF30-P-3 TF30-P-103	Translating Cowl and Splitter Plate Triple Plow I	Short	Light Weight
EF-111A	TF30-P-3 TF30-P-103 TF30-P-109	Translating Cowl and Splitter Plate Triple Plow I	Short	Light Weight
F-111B	TF30-P-12	First 5 Aircraft – Translating Cowl and Splitter Plate, Last 2 and Production Design – Blow-in Doors	Long	Heavy Weight with a Catapult Launch Bar Attached To The Nose Gear and a powered Tail Arresting Hook
F-111C (Royal Australian Air Force Version)	TF30-P-3 TF30-P-103 TF30-P-109	Translating Cowl and Splitter Plate Triple Plow I	Long	Heavy Weight
F-111D	TF30-P-9 TF30-P-109	Blow-In Doors Triple Plow II	Short	Light Weight
F-111E	TF30-P-3 TF30-P-103 TF30-P-109	Blow-In Doors Triple Plow II	Short	Light Weight
F-111F	TF30-P-100 TF30-P-109	Blow-In Doors Triple Plow II	Short	Light Weight
F-111G (Ex-FB-111As)	TF30-P-107 TF30-P-108 (RAAF hybrid)	Blow-In Doors Triple Plow II	Long	Heavy Weight
F-111K British (Royal Air Force Version)	TF30-P-3	Translating Cowl and Splitter Plate Triple Plow I	Long	Heavy Weight
FB-111A (SAC Version)	TF30-P-7 TF30-P-107	Blow-In Doors Triple Plow II	Long	Heavy Weight

APPENDIX 3: AVIONICS SYSTEMS

Model	Bombing and Navigation	TFR System	Radar System
F-111A	AN/AJQ-20A (Analog System)	AN/APQ-110	AN/APQ-113
EF-111A	AN/AJQ-20A (Analog System)	AN/APQ-110	AN/APQ-113
F-111B	AN/AJN-14 Inertial Navigation System	None	AN/AWG-9 (AIM-54A Phoenix Missile Firepower Control System)
F-111C	AN/AJQ-20A (Analog System) AUP Avionics (Digital System)	AN/APQ-110 AN/APQ-171	AN/APQ-113 AN/APQ-169
F-111D	Mark II Avionics (Digital System)	AN/APQ-128	AN/APQ-130
F-111E	AN/AJQ-20A (Analog System)	AN/APQ-110	AN/APQ-113
F-111F	AN/AJN-16 (Digital System)	AN/APQ-146 or AN/APQ-171	AN/APQ-161 or AN/APQ-169
F-111G	Mark IIB Avionics (Digital System) AMP (Digital System)	AN/APQ-134 AN/APQ-171	AN/APQ-114 AN/APQ-169
FB-111A	Mark IIB Avionics (Digital System)	AN/APQ-134	AN/APQ-1114

APPENDIX 4: F-111 SPECIFICATIONS

	Length	Wing Span Ext/Swpt	Height	Empty Weight	Maximum Weight	Maximum Speed
F-111A	73' 6"	63' 0"/31' 11"	17' 1.4"	45,200 LBS	92,500 LBS	Mach 2.3
EF-111A	73' 6"	63' 0"/31' 11"	17' 1.4"	53,600 LBS	87,800 LBS	Mach 2.2
F-111B (1-5)	66' 9"	70' 0"/33'11"	15' 9"	46,112 LBS	79,000 LBS	Mach 2.2
F-111B (6+7)	68' 9"	70' 0"/33'11"	15' 9"	46,112 LBS	79,000 LBS	Mach 2.2
F-111C	73' 6"	70' 0"/33' 11"	17' 1.4"	50,000 LBS	114,300 LBS	Mach 2.5
F-111D	73' 6"	63' 0"/31' 11"	17' 1.4"	46,900 LBS	92,500 LBS	Mach 2.5
F-111E	73' 6"	63' 0"/31' 11"	17' 1.4"	45,700 LBS	92,500 LBS	Mach 2.5
F-111F	73' 6"	63' 0"/31' 11"	17' 1.4"	47,200 LBS	100,000 LBS	Mach 2.5
F-111G	73' 6"	70' 0"/33' 11"	17' 1.4"	49,000 LBS	114,300 LBS	Mach 2.5
FB-111A	73' 6"	70' 0"/33' 11"	17' 1.4"	45,200 LBS	92,500 LBS	Mach 2.2

APPENDIX 5: F-111 ENGINE SPECIFICATIONS

Designation		Weight Thrust	Military Thrust In A/B	Maximum
F-111A PRE-PRODUCTION				
TF30-P-1	JTF-10A-20	4,100 LBS	10,750	18,500 LBS
F-111A, EF-111A F-111C, F-111E				
TF30-P-3	JTF-10A-21	4,062 LBS	10,750	18,500 LBS
NAVY F-111B				
TF30-P-12	JTF-10A-27A	4,000 LBS	12,290	20,000 LBS
F-111D				
TF30-P-9	JTF-10A-36	4,070 LBS	12,000	19,600 LBS
FB-111A				
TF30-P-7	JTF-10A-27D	4,121 LBS	12,290	20,250 LBS
F-111F				
TF30-P-100	JTF-10A-32C	3,900 LBS	14,560	25,100 LBS

APPENDIX 6: F-111 UNIT AND TAIL CODE SUMMRY

USAF TACTICAL F-111 UNITS

20th TFW/FW/OG, RAF Upper Heyford, England

UR	79th TFS	Yellow	10/70 thru 1/71
US	55th TFS		Not used
UT	77th TFS	Red	12/70 thru 1/71
JR	79th TFS	Yellow	1/71 thru 11/72
JS	55th TFS	Blue	4/71 thru 11/72
JT	77th TFS	Red	1/71 thru 11/72
UH	20th TFW/FW/OG Wing Code		

42nd ECS		2/3/84 thru 1/25/91 66th ECW, then to 20th TFW thru 7/1/92
55th TFS/FS	Blue	8/7/72 thru 10/1/91 – reassigned to 20th OG until 12/30/93
77th TFS/FS	Red	7/72 thru 10/1/91 – reassigned to 20th OG until 12/30/93
79th TFS/FS	Yellow	8/7/72 thru 10/1/91 – reassigned to 20th OG until 12/30/93

27th TFW/FW, Cannon AFB, New Mexico

CA	481st TFS	Green	F-111A 10/69 thru 1/71
			F-111E 9/30/69 thru 7/71
CC	522nd TFS	Red	F-111A 1971
			F-111D 5/72 thru 12/72
			F-111E 10/71 thru 11/71
CC	Det 2 57th FWW		10/01/70 thru 5/1/72
CD	524th TFS	Yellow	F-111A 1972
			F-111D 1972
			F-111E 1972
CE	4427 TFRS	Purple	F-111D 11/1/71 thru 1972
CC	27th TFW/FW Wing Code		

428th TFS/TFTS/FS	Blue
429th ECS	Black with Black Falcon on tail cap antenna fairing
430th ECS	Black
481st TFTS	Green
522nd TFS/FS	Red
523rd TFS/FS	Blue
524th TFS/TFTS/FS	Yellow
4427th TFRS	Purple

48th TFW/FW RAF Lakenheath, England

LN	48th TFW/FW Wing Tail Code		

492nd TFS/FS	Blue	F-111F 4/22/77 thru 1992 (converted to F-15E)
493rd TFS/FS	Yellow	F-111F 4/22/77 thru 12/18/92 (inactivated – activated with F-15C/D 1/1/94)
494th TFS/FS	Red	F-111F 4/22/77 thru 1992 (converted to F-15E)
495th TFS/FS	Green	F-111F 4/22/77 thru 12/31/91 inactivated

57th FWW/FW/Wing Nellis AFB, Nevada

WF	4539th CCTS	7/10/68 thru 10/15/69
	422nd FWS	10/15/69 thru 10/1/71
WA	57th FWW/FW/Wing Wing Tail Code	
	422nd FWS/TES	F-111F 4/22/77 thru 1992 (converted to F-15E)
	431st FWS/TES	F-111F 4/22/77 thru 12/18/92 (inactivated – activated with F-15C/D 1/1/94)

347th TFW Mountain Home, Idaho

MO	366th TFW/FW Wing Tail Code	
	391st TFS	F-111F 5/15/71 thru 10/30/72
	4589th TFS	F-111F 9/1/71 thru 10/30/72
	4590th TFS	F-111F 1/1/72 thru 10/30/72

347th TFW, Korat RTAFB, Thailand

HG	347th TFW Wing Tail Code	
	428th TFS	F-111A
	429th TFS	F-111A
	430th TFS	F-111A

366th TFW/FW Mountain Home, Idaho

MO	366th TFW/FW Wing Tail Code		
	388th TFTS	Yellow	F-111A
		Red	EF-111A
	389th TFS/TFTS	Red	F-111F thru 4/22/77
		Yellow	F-111A
	390th TFS	Green	F-111F thru 4/22/77
			F-111A
	390th ECS		EF-111A
	391st TFS	Blue	F-111F thru 4/22/77
			F-111A
	429th ECS	Black	EF-111A

474th TFW Nellis AFB, Nevada

NA	428th TFS	Blue	F-111A
NB	429th TFS	Yellow	F-111A
NC	430th TFS	Red	F-111A
ND	442nd TFTS	Green	F-111A/F-111E
NA	474th TFW Wing Tail Code		
	428th TFS	Blue	F-111A
	429th TFS	Yellow	F-111A
	430th TFS	Red	F-111A
	442nd TFTS	Green	F-111A

STRATEGIC AIR COMMAND F-111 UNITS

340th Bomb Group, Carswell AFB, Texas

    4111th BS/9th BS   FB-111A

380th BW (Medium) Plattsburgh AFB, New York

528th BS (M)	Yellow (later Blue)	FB-111A
529th BS (M)	Red	FB-111A
4007th/530th	CCTS	FB-111A

509th BW (Medium) Pease AFB, New Hampshire

    393rd BS (M)
    715th BS (M)

USAF TEST F-111 UNITS

46th Test Wing, Eglin AFB, Florida
(3246th Test Wing)

AD & ET   3247th Test Squadron   Red diamonds on a White fin band

6510th Test Wing Edwards AFB, California

ED   6512th Test Squadron   White Xs on a Blue fin band
         431th Test Squadron

USAF Air Warfare Center, Eglin AFB, Florida

OT   4485th Test Squadron   Black & White checkerboard tail stripe

Sacramento Air Logistics Center (SM-ALC), McClellan AFB, California

SM   2874th Test Squadron Engineering Flight Test triangle on the tail above the serial number
        337th Test Squadron

ROYAL AUSTRALIAN AIR FORCE F-111 UNITS

No. 82 Wing RAAF Base, Amberley

    No. 1 Squadron RAAF (Blue lightning bolt)
    No. 6 Squadron RAAF (Yellow lightning bolt or large No 1)
    ADRU (during the 1980s permanently operated A8-132)

APPENDIX 7: ATTRITTED F-111 AIRCRAFT

	MODEL	GD #	TAIL #	DATE	WHERE
1.	F-111A	9	63-9774	JAN 19, 1967	EDWARDS AFB
2.	F-111B	4	151973	APR 21, 1967	BETHPAGE NY
3.	F-111A	15	63-9780	OCT 19, 1967	BOYD TX
4.	F-111A	19	65-5701	JAN 2, 1968	EDWARDS AFB
5.	F-111A	40	66-0022	MAR 28, 1968	N. VIETNAM
6.	F-111A	35	66-0017	MAR 30, 1968	N. VIETNAM
7.	F-111A	42	66-0024	APR 21, 1968	N. VIETNAM
8.	F-111A	50	66-0032	MAY 8, 1968	NELLIS AFB
9.	F-111A	4	63-9769	MAY 18, 1968	HOLLOMAN AFB, NEW MEXICO
10.	F-111B	2	151971	SEP 11, 1968	PACIFIC OCEAN OFF CA. COAST
11.	F-111A	58	66-0040	SEP 23, 1968	NELLIS AFB
12.	F-111A	60	66-0042	FEB 12, 1969	NELLIS AFB
13.	F-111A	61	66-0043	MAR 4, 1969	NELLIS AFB
14.	F-111A	88	67-0043	MAY 22, 1969	TUBA CITY, ARIZONA
15.	F-111A	94	67-0049	DEC 22, 1969	NELLIS AFB
16.	FB-111A	25	68-0253	OCT 7, 1970	CARSWELL AFB
17.	FB-111A	55	68-0283	JAN 8, 1971	LAKE PONCHATRAIN, LOUISIANA
18.	F-111E	3	67-0117	APR 23, 1971	EDWARDS AFB
19.	F-111A	47	66-0029	SEP 1, 1971	CANNON AFB
20.	F-111E	28	68-0018	JAN 18, 1972	RAF UPPER HEYFORD
21.	F-111F	46	70-2407	FEB 2, 1972	GD/FW RUN STATION #3
22.	F-111A	81	67-0036	APR 24, 1972	GRAND CANYON, ARIZONA
23.	F-111F	49	70-2410	JUN 15, 1972	NEAR FALLON, NEVADA
24.	F-111A	127	67-0082	JUN 18, 1972	EGLIN AFB
25.	F-111A	21	65-5703	SEP 11, 1972	EDWARDS AFB
26.	F-111A	123	67-0078	SEP 28, 1972	N. VIETNAM
27.	F-111A	111	67-0066	OCT 16, 1972	N. VIETNAM
28.	F-111A	108	67-0063	NOV 6, 1972	N. VIETNAM
29.	F-111A	137	67-0092	NOV 20, 1972	N. VIETNAM
30.	F-111A	144	67-0099	DEC 18, 1972	N. VIETNAM
31.	F-111A	113	67-0068	DEC 22, 1972	N. VIETNAM
32.	F-111E	34	68-0024	JAN 11, 1973	RAF UPPER HEYFORD
33.	F-111A	117	67-0072	FEB 21, 1973	TAKHLI RTAFB, THAILAND
34.	F-111D	21	68-0105	MAR 20, 1973	HOLBROOK, ARIZONA
35.	F-111D	74	68-0158	MAR 20, 1973	HOLBROOK, ARIZONA
36.	F-111E	18	68-0008	MAY 15, 1973	SCOTLAND
37.	F-111A	156	67-0111	JUN 16, 1973	CAMBODIA
38.	F-111A	85	67-0040	JUL 11, 1973	UTAH
39.	F-111D	29	68-0113	DEC 21, 1973	NEW MEXICO
40.	F-111F	34	70-2395	SEP 11, 1974	NEW MEXICO
41.	F-111A	100	67-0055	NOV 12, 1974	KINGSTON, UTAH
42.	FB-111A	52	68-0280	FEB 3, 1975	VERMONT
43.	FB-111A	67	69-6505	FEB 3, 1975	VERMONT
44.	F-111E	91	68-0081	MAR 5, 1975	SCOTLAND
45.	F-111A	52	66-0034	JUN 6, 1975	ARIZONA
46.	F-111A	43	66-0025	JUN 20, 1975	NELLIS AFB
47.	F-111A	76	66-0058	OCT 7, 1975	NELLIS AFB
48.	F-111E	70	68-0060	NOV 5, 1975	RAF UPPER HEYFORD
49.	F-111F	32	70-2393	NOV 8, 1975	MOUNTAIN HOME AFB
50.	FB-111A	62	68-0290	DEC 23, 1975	MAINE

MODEL	GD #	TAIL #	DATE	WHERE
51. F-111A	125	67-0080	MAR 11, 1976	NELLIS AFB
52. F-111F	27	78-2388	MAR 16, 1976	MOUNTAIN HOME AFB
53. F-111A	105	67-0060	APR 7, 1976	WENDOVER RANGE, UTAH
54. FB-111A	73	69-6511	JUL 7, 1976	VERMONT
55. F-111D	83	68-0167	OCT 10, 1976	ROSWELL, NEW MEXICO
56. F-111E	2	67-0116	OCT 27, 1976	EGLIN AFB
57. FB-111A	38	68-0266	FEB 14, 1977	NEW HAMPSHIRE
58. F-111F	85	73-00709	APR 21, 1977	NAS CHINA LAKE
59. F-111C	12	A8-136	APR 28, 1977	AUSTRALIA
60. F-111D	62	68-0146	SEP 2, 1977	NEW MEXICO
61. F-111C	9	A8-133	SEP 29, 1977	NEW SOUTH WALES
62. F-111D	9	68-0093	OCT 3, 1977	NEW MEXICO
63. F-111F	94	73-00718	OCT 5, 1977	NORTH OF RAMSTEIN AB, GERMANY
64. FB-111A	57	68-0285	OCT 28, 1977	MAINE
65. F-111E	80	68-0070	OCT 31, 1977	WALES
66. F-111A	128	67-0083	NOV 30, 1977	NEVADA
67. F-111F	19	70-2380	DEC 15, 1977	ENGLAND
68. F-111F	93	73-00717	MAR 29, 1978	ENGLAND
69. F-111C	17	A8-141	OCT 25, 1978	NEW ZEALAND
70. F-111D	89	68-0173	NOV 18, 1978	ARIZONA
71. F-111A	104	67-0059	JAN 4, 1979	MOUNTAIN HOME AFB
72. F-111D	25	68-0109	FEB 16, 1979	NEW MEXICO
73. F-111F	6	70-2367	APR 20, 1979	SCOTLAND (MIDAIR)
74. F-111F	90	73-00714	APR 20, 1979	SCOTLAND (MIDAIR)
75. F-111A	150	67-0105	JUL 5, 1979	NELLIS AFB RED FLAG
76. F-111E	52	68-0042	JUL 24, 1979	ENGLAND
77. F-111A	70	66-0052	JUL 31, 1979	MOUNTAIN HOME AFB
78. F-111C	13	A8-137	AUG 24, 1979	NEW ZEALAND
79. FB-111A	33	68-0261	SEP 18, 1979	NELLIS AFB RED FLAG
80. F-111E	22	68-0012	OCT 30, 1979	ENGLAND
81. F-111E	55	68-0045	DEC 12, 1979	ENGLAND
82. F-111E	13	68-0003	DEC 19, 1979	SCOTLAND
83. F-111D	35	68-0119	FEB 6, 1980	NEW MEXICO
84. F-111A	142	67-0097	MAR 26, 1980	OWHYHEE COUNTY, IDAHO
85. F-111E	67	68-0057	APR 29, 1980	ENGLAND
86. F-111D	55	68-0139	JUL 14, 1980	9 MILES NORTH OF CANNON AFB
87. FB-111A	51	68-0279	JUL 30, 1980	CANADA
88. FB-111A	40	68-0268	OCT 6, 1980	JONESPORT, MAINE
89. FB-111A	35	68-0263	JAN 30, 1981	PORTSMOUTH, NEW HAMPSHIRE
90. F-111F	71	72-1441	FEB 4, 1981	ENGLAND
91. F-111A	118	67-0073	JAN 19, 1982	MOUNTAIN HOME AFB
92. F-111D	26	68-0110	JAN 27, 1982	SACRAMENTO, CA.
93. F-111A	63	66-0045	MAY 12, 1982	MOUNTAIN HOME AFB
94. F-111F	77	72-1447	JUN 23, 1982	RAF LAKENHEATH
95. F-111D	76	68-0160	SEP 14, 1982	NEW MEXICO
96. F-111F	97	74-00179	SEP 16, 1982	SCOTLAND
97. F-111A	143	67-0098	OCT 8, 1982	MOUNTAIN HOME AFB
98. F-111F	92	73-00716	NOV 1, 1982	KONYA RANGE, TURKEY
99. F-111A	138	67-0093	NOV 9, 1982	MOUNTAIN HOME AFB, SAC PAD
100. F-111F	16	70-2377	DEC 7, 1982	ISLE OF SKYE, SCOTLAND
101. F-111A	72	66-0054	APR 13, 1983	SAYLOR CREEK MOUNTAIN HOME AFB
102. F-111F	106	74-00188	APR 26, 1983	BORKUM ISLAND, HOPSTEN, GERMANY
103. FB-111A	14	68-0242	JUN 7, 1983	ARIZONA (RED FLAG)
104. F-111F	5	70-2366	DEC 21, 1983	NORTH SEA, SCARBOUROGH UK

MODEL	GD #	TAIL #	DATE	WHERE
105. F-111A	44	66-0026	MAR 13, 1984	SAYLOR CREEK, IDAHO
106. F-111E	29	68-0019	AUG 9, 1984	KINLOSS, SCOTLAND
107. F-111D	80	68-0164	OCT 17, 1984	HOLLOMAN AFB, NEW MEXICO
108. F-111C	15	A8-139	JAN 28, 1986	AUSTRALIA
109. F-111F	28	70-2389	APR 15, 1986	TRIPOLI, LIBYA
110. F-111F	57	70-2418	FEB 23, 1987	NEWMARKET, UK
111. F-111C	4	A8-128	APR 2, 1987	TENNERFIELD, AUSTRALIA
112. F-111F	14	70-2375	JUL 28, 1987	SCOTLAND
113. F-111D	46	68-0125	SEP 11, 1987	9 MILES N OF CANNON AFB
114. F-111A	147	67-0102	JAN 2, 1988	MOUNTAIN HOME AFB, END OF RUNWAY
115. F-111D	48	68-0132	MAR 17, 1988	CANNON AFB
116 F-111D	14	68-0098	JUN 8, 1988	MELROSE RANGE
117. F-111D	46	68-0130	OCT 21, 1988	CANNON AFB
118. FB-111A	15	68-0243	FEB 2, 1989	ST JOHNSBURG, VERMONT
119. F-111F	36	70-2397	APR 5, 1989	NELLIS AFB RED FLAG RANGE
120. F-111E	11	68-0001	FEB 5, 1990	THE WASH, UK
121. F-111D	84	68-0168	MAR 26, 1990	BIRDSTRIKE/WRITTEN OFF
122. F-111F	7	70-2368	MAY 2, 1990	SCULTHORPE, UK
123. F-111E	76	68-0066	JUL 20, 1990	INCIRLIK AB, TURKEY
124. F-111D	47	68-0131	AUG 23, 1990	HOLLOMAN AFB
125. F-111F	101	74-0183	OCT 10, 1990	SAUDI ARABIA
126. EF-111A	41	66-0023	FEB 14, 1991	SAUDI ARABIA
127. F-111D	56	68-0140	1990-1993	WRITE-OFF
128. F-111D	49	68-0133	1990-1993	WRITE-OFF
129. EF-111A	74	66-0056	FEB 4, 1992	FINMERE, UK
130. F-111E	62	68-0052	SEP 17, 1992	RAF UPPER HEYFORD
131. F-111C	3	A8-127	SEP 13, 1993	GUYRA, NEW S. WALES
132. F-111F	93	70-2412	SEP 22, 1993	MELROSE RANGE, CANNON AFB
133. F-111F	9	70-2370	JULY 1994	FIRE WRITE-OFF
134. F-111F	47	70-2408	1994-1995	WRITE-OFF
135. F-111F	3	70-2364	1994-1995	WRITE-OFF
136. F-111E	50	68-0040	FEB 16, 1995	CANNON AFB, END OF RUNWAY
137. EF-111A	62	66-0044	JUN 17, 1996	TUCUMCARI, NEW MEXICO

Numbers 121, 127, 128, 129, 133, 134, 135, 136 were accidents, not crashes. The airframe was written off.

APPENDIX 8: F-111 AIRCREW FATALITIES WORLDWIDE 1967-1998
Total number of crewmembers killed – 113

Date	Name	Tail #	Model/Callsign
January 19, 1967	Maj Herbert F. Brightwell	63-9774	F-111A
April 21, 1967	Mr. Ralph H. Donnell Mr. Charles E. Wangeman	151973	F-111B
March 27, 1968	Maj Henry McCann Capt Dennis Graham	66-0022	F-111A Omaha 77
April 22, 1968	Cmdr David L. Cooley USN Lt Col Edwin D. Palmgren	66-0024	F-111A Tailbone 78
September 11, 1968	Mr. Barton Warren Mr. Anthony Byland	151971	F-111B
February 12, 1969	Capt Robert E. Jobe Capt William D. Fuchlow	66-0042	F-111A
December 22, 1969	Lt Col Thomas J. Mack Maj James L. Anthony	67-0049	F-111A
October 7, 1970	Lt Col Robert S. Montgomery Lt Col Charles G. Robinson	68-0253	FB-111A
January 8, 1971	Lt Col Bruce D. Stocks Maj Billy C. Gentry	68-0283	FB-111A
April 23, 1971	Maj James W. Hurt III Maj Robert J. Furman	67-0117	F-111E
January 18, 1972	Lt Col Floyd B. Sweet Lt Col Kenneth T. Black	68-0018	F-111E Sewn 11
June 18, 1972	Col Keith E. Brown Lt Col James D. Blank	67-0082	F-111A Flick 79
September 28, 1972	Maj William C. Coltman Lt Robert A. Brett Jr.	67-0078	F-111A Ranger 23
October 17, 1972	Capt James A. Hockridge Lt Allen U. Graham	67-0066	F-111A Coach 33
November 7, 1972	Maj Robert Brown Capt Robert D. Morrissey	67-0063	F-111A Whaler 57
November 20, 1972	Capt Donald D. Stafford Capt Charles J. Cafferrelli	67-0092	F-111A Burger 54
December 18, 1972	Lt Col Ronald J. Ward Maj James R. McElvain	67-0099	F-111A Snug 40

Date	Name	Tail #	Model/Callsign
March 20, 1973	Maj William W. Gude Capt David C. Blackledge	68-0105	F-111D
March 20, 1973	Maj Richard L. Brehm Capt William T. Halloran	68-0158	F-111D
December 21, 1973	Capt William K. Delaplane III Lt Robert J. Kierce	68-0113	F-111D
September 11, 1974	Capt William A. Kennedy Capt David C. McKennon	70-2395	F-111F
October 7, 1975	Capt Ralph "Dave" Bowles Maj Merle D. Kenney	66-0058	F-111A
February 14, 1977	Capt Edward A. Riley Capt Jeremiah F. Sheehan	68-0266	FB-111A
March 29, 1977	Sq Ldr John Holt RAAF Flt Lt Phil Noordink RAAF	A8-133	F-111C
October 3, 1977	Capt Richard L. Cardenas Capt Steven G. Nelson	68-0093	F-111D
October 5, 1977	Capt Stephen H. Reid Capt Carl T. Poole	73-0718	F-111F
October 31, 1977	Capt John J. Sweeney Capt William W. Smart	68-0070	F-111E
November 30, 1977	Capt Arthur Stowe Maj Lorley Wagner	67-0083	F-111A
March 29, 1978	Capt Charles H. Kitchell 1Lt Jeffrey T. Moore	73-0717	F-111F
July 5, 1979	Maj Gary A. Mekash Lt Col Eugene H. Soeder	67-0105	F-111A
July 14, 1979	Capt David W. Powell Capt Douglas A. Pearce	68-0042	F-111E
July 31, 1979	2Lt Larry E. McFarland Capt Myles D. Hammon	66-0052	F-111A
September 8, 1979	Capt Phillip B. Donovan Capt William J. Full	68-0261	FB-111A
December 12, 1979	Capt Randolph P. Gaspard Maj Frank B. Slusher	68-0045	F-111E LAY 40
December 19, 1979	Capt Richard A. Hetzner Capt Raymond C. Spaulding	68-0003	F-111E

Date	Name	Tail #	Model/Callsign
February 6, 1980	Capt Roy W. Westerfield 2Lt Steven P. Anderson	68-0119	F-111D
April 29, 1980	Capt Jack A. Hines Capt Richard J. Franks	68-0057	F-111E
March 28, 1980	Capt Joseph G. Raker Capt Larry R. Honza	67-0097	F-111A
July 14, 1980	Maj Ulysses S. Taylor III 1Lt Paul E. Yeager	68-0139	F-111D
October 6, 1980	Maj Thomas M. Mullen Capt Gary A. Davis	68-0268	FB-111A
September 14, 1982	Maj Howard L. Tallman III Capt William R. Davy	68-0160	F-111D
December 7, 1982	Maj Burnley L. Rudiger 1Lt Steven J. Pitt	70-2377	F-111F
April 26, 1983	Capt Charles M. Vidas 1Lt Steven A. Groark	74-0188	F-111F
March 13, 1984	Capt Steven F. Locke Capt David K. Peth	66-0026	F-111A
October 17, 1984	Capt Alan J. Pryor 1Lt Albert H. Torn	68-0164	F-111D
January 28, 1986	Flt Lt Stephen Erskine RAAF Capt Gregory S. Angell USAF	A8-139	F-111C
April 15, 1986	Maj Fernando L. Ribas-Dominicci Capt Paul F. Lorence	70-2389	F-111F Karma 52
April 3, 1987	Flt Lt Mark Fallon RAAF Flt Off. William Pike RAAF	A8-128	F-111C Buckshot 01
July 28, 1987	Capt Thomas F. "Chip" Stem Capt Philip D. "Phil" Baldwin	70-2375	F-111F Badger 21
January 12, 1988	Capt Robert A. Meyer Jr. Capt Frederick A. Gerhart	67-0102	F-111A Honest 21
June 8, 1988	Capt Glenn E. Troster Capt Michael A. Barritt	68-0098	F-111D Cubid 12
April 5 1989	1Lt Bob Boland Capt James A. Gleason	70-2397	F-111F Greebie 54
February 5, 1990	Capt Clifford W. Massengill 1Lt Thomas G. Dorsett	68-0001	F-111E

APPENDICES

Date	Name	Tail #	Model/Callsign
October 10, 1990	Capt Frederick "Art" Reid Capt Thomas R. "TC" Caldwell	74-0183	F-111F Cougar 41
February 14, 1991	Capt Douglas M. Bradt Capt Paul R. Eichenlaub II	66-0023	EF-111A Ratchet 75
September 17, 1992	Capt Jerry C. Lindh Capt David Michael McGuire	68-0052	F-111E
September 13, 1993	Flt Lt Jeremy McNess RAAF Flt Lt Mark Cairns-Cowan RAAF	A8-127	F-111C

APPENDIX 9: F-111 AIRCRAFT ON DISPLAY

M/D/S	S/N	LOCATION
F-111A	63-9766	EDWARDS AFB
F-111A	63-9767	OCTAVE-CHANUTE AEROSPACE MUSEUM
F-111A	63-9771	CANNON AFB, 27 FW
F-111A	63-9773	SHEPPARD AFB, 82 TRW
F-111A	63-9775	U.S. SPACE AND ROCKET CENTER
RF-111A	63-9776	MOUNTAIN HOME AFB, 366 WG
F-111A	63-9778	EDWARDS AFB
F-111A	67-0067	WRIGHT-PATTERSON AFB, USAF MUSEUM
F-111A	67-0100	NELLIS AFB
F-111E	67-0120	IMPERIAL WAR MUSEUM, DUXFORD
FB-111A	67-0159	MCCLELLAN AFB
F-111E	68-0011	RAF LAKENHEATH
F-111E	68-0020	HILL AFB
F-111E	68-0033	PIMA AIR MUSEUM
F-111E	68-0055	ROBINS AFB
F-111E	68-0058	EGLIN AFB
FB-111A	68-0245	MARCH FIELD MUSEUM FOUNDATION
FB-111A	68-0248	ELLSWORTH AFB, 28 BW
FB-111A	68-0267	STRATEGIC AIR COMMAND MUSEUM
FB-111A	68-0275	KELLY AFB, SA-ALC
FB-111A	68-0284	BARKSDALE AFB, 2 BW
FB-111A	68-0286	PLATTSBURGH AFB
FB-111A	69-6507	CASTLE AIR MUSEUM
FB-111A	69-6509	WHITEMAN AFB, 509 BW
F-111F	70-2390	WRIGHT-PATTERSON AFB, USAF MUSEUM
F-111F	74-0177	AMARC
F-111F	74-0178	AMARC
F-111F	74-0187	AMARC

NOTES

NOTES

NOTES

NOTES

Also from the Publisher

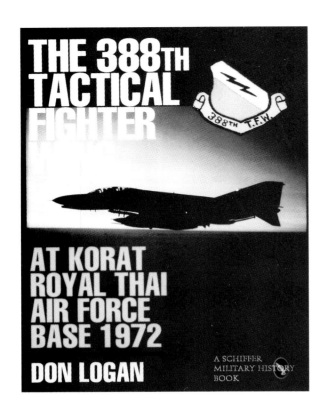
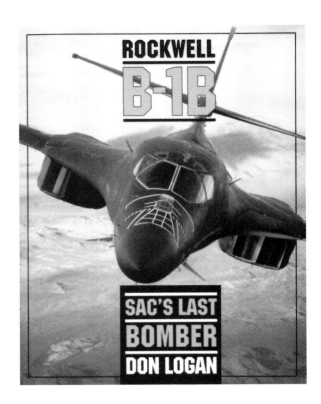
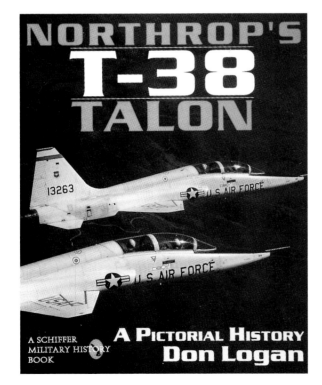
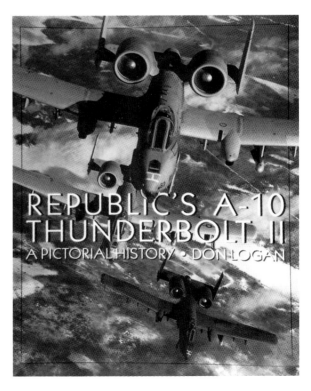